Instrumentation and Measurement Methods for Industry 4.0 and IoT

Instrumentation and Measurement Methods for Industry 4.0 and IoT

Guest Editor

Mauro Serpelloni

MDPI

Basel • Beijing • Wuhan • Barcelona • Belgrade • Novi Sad • Cluj • Manchester

Guest Editor
Mauro Serpelloni
Department of Information
Engineering
University of Brescia
Brescia
Italy

Editorial Office
MDPI AG
Grosspeteranlage 5
4052 Basel, Switzerland

This is a reprint of the Special Issue, published open access by the journal *Instruments* (ISSN 2410-390X), freely accessible at: https://www.mdpi.com/journal/instruments/special_issues/909UG5F4AX.

For citation purposes, cite each article independently as indicated on the article page online and as indicated below:

Lastname, A.A.; Lastname, B.B. Article Title. *Journal Name* **Year**, *Volume Number*, Page Range.

ISBN 978-3-7258-8161-1 (Hbk)
ISBN 978-3-7258-8162-8 (PDF)
https://doi.org/10.3390/books978-3-7258-8162-8

Contents

About the Editor

Mauro Serpelloni

Mauro Serpelloni is a Full Professor at the University of Brescia, Department of Information Engineering, working in the field of Electrical and Electronic Measurements. His research focuses on smart sensing systems, IoT-enabled measurement platforms, wearable and printed sensors, wireless and autonomous sensing systems, energy harvesting, and cyber-physical systems for industrial and biomedical applications. He has coordinated and participated in several national and international research projects, including PRIN, Horizon Europe, PON, and regional Industry 4.0 initiatives, with activities related to sensing infrastructures, smart devices, and digital technologies. He is an IEEE Senior Member, Coordinator of the PhD Programme in Technology for Health at the University of Brescia, and General Chair of IEEE MetroInd4.0&IoT 2023 and 2025.

instruments MDPI

Editorial

Instrumentation and Measurement Methods for Industry 4.0 and IoT

Mauro Serpelloni

Department of Information Engineering, University of Brescia, 25123 Brescia, Italy; mauro.serpelloni@unibs.it

The current industrial transformation is changing the way production systems are designed, manufactured, monitored, controlled, and maintained. Compared to traditional automation, this transformation is characterized by pervasive connectivity, cyber-physical integration, distributed intelligence, and the seamless exchange of information between physical assets and digital services. In this scenario, the Internet of Things (IoT), artificial intelligence, cloud and edge computing, digital twins, advanced sensors, and data analytics are no longer independent technologies. They are progressively becoming elements of a single technological ecosystem where measurement data represents the essential interface between the physical and digital worlds [1]. The COVID-19 pandemic has accelerated several aspects of this transition, making remote working, telemedicine, online services, distributed monitoring, and digital communication essential tools for business continuity and social resilience. This acceleration has highlighted the need for always-connected, remotely accessible, and data-driven systems. Such systems require the ability to observe, measure, transmit, and interpret physical phenomena remotely. For this reason, the development of new sensors, acquisition techniques, data processing methods, and distributed measurement architectures has become increasingly important. This transition has profound consequences for metrology. Industry 4.0 can only create value if decisions are based on reliable, traceable, and timely information. Smart factories, in-line inspection, predictive maintenance, remote monitoring, and real-time control require measurement systems capable of operating in complex environments, often under variable operating conditions and with limited human supervision. The concept of Metrology 4.0 summarizes this evolution, emphasizing the digitalization of metrological services, the integration of measuring instruments into production systems, and the need for real-time calibration, process monitoring, traceability, and quality control [2]. From this perspective, metrology is not simply a production support activity, but an enabling infrastructure for smart, sustainable, and resilient manufacturing. The adoption of Metrology 4.0 requires qualified personnel, adequate infrastructure, management commitment, awareness of Industry 4.0 technologies, investment capacity, flexible organizational culture, and appropriate standards and norms [2]. Recent work on smart coordinate measuring machines confirms this trend, demonstrating that dimensional metrology is moving from isolated inspection tools towards connected, multi-sensor, and uncertainty-aware components of cyber-physical manufacturing systems [3]. Similar needs are emerging in additive manufacturing, where smart metrology combines in situ sensing and machine learning-based analytics to support defect detection, process optimization, predictive quality monitoring, and future closed-loop control [4]. Modern industrial and IoT scenarios require measurement systems that are not only accurate but also connected, scalable, interoperable, and able to operate under complex and variable conditions. Sensors, embedded electronics, wireless networks, edge and cloud computing, artificial intelligence, and advanced signal processing techniques are increasingly integrated into measurement chains. This convergence opens up new opportunities for real-time monitoring, diagnostics, localization, tracking, virtual measurements, human-centric applications, and smart devices. At the same time, it raises challenges related to calibration, uncertainty assessment, reliability, cybersecurity, energy efficiency, data governance, and long-term stability.

Digital twins represent one of the most prominent examples of this convergence. Their reliability depends on the continuous alignment between the physical system and its virtual counterpart. This alignment requires calibrated sensors, validated models, synchronization strategies, and methods for managing measurement uncertainty. Recent studies have highlighted the importance of calibration methods based on computer vision, event-based sensing, and neuromorphic vision to improve the accuracy of digital twins in cyber-physical manufacturing systems [5]. Verification and validation are equally important, as a digital twin can only support decisions when its credibility, fidelity, and domain validity are clearly assessed [6].

Artificial intelligence has become one of the most important tools for extracting actionable insights from industrial data. AI can support anomaly detection, fault diagnosis, production planning, quality control, adaptive monitoring, and maintenance optimization [7]. However, its adoption in Industry 4.0 is still limited by issues related to system integration, heterogeneous data, workforce factors, and AI reliability [1]. From a measurement perspective, these issues must also be considered as metrological problems. A forecast cannot be fully useful if its uncertainty is unknown, if its outcome cannot be explained, or if the data acquisition process is not reproducible.

Predictive maintenance is another central topic of Industry 4.0. In [8], for example, signals acquired by sensors are processed, and deep learning architectures are used for fault classification, while uncertainty quantification and SHAP-based interpretability provide confidence and interpretability. This approach is significant because it combines data-driven diagnosis with metrological awareness: the classification result is not treated simply as a black-box output, but as information whose reliability and physical meaning must be assessed. Other studies confirm the importance of scalable architectures for predictive maintenance. Edge-cloud and edge-fog-cloud approaches can reduce latency, distribute computational load, and support real-time decision making in resource-constrained environments [9,10]. These frameworks demonstrate that predictive maintenance is not only an algorithmic problem, but also a measurement system problem involving sensor placement, data acquisition, heterogeneous data streams, model implementation, and feedback to operators.

Security is also part of this picture. Cybersecurity, privacy, access control, and transparency in decision-making are closely related to measurement integrity. Recent studies on explainable AI for IoT cybersecurity emphasize the importance of transparent, auditable, and adaptive methods for anomaly detection and decision support in large-scale connected infrastructures [11]. This is particularly relevant when measurement data is used to guide safety-critical or economically relevant decisions.

In human- and society-centric applications, this new concept of Metrology 4.0 is emerging even more rapidly. Recent human–computer interaction systems show a development towards radar, vision, haptic, wearable, and wireless sensor technologies capable of obtaining robust information about users, environments, and activities [12]. In these systems, sensor fusion is essential because no single modality can provide complete and reliable information under all operating conditions. Healthcare and rehabilitation represent particularly important application fields. IoT and Internet of Medical Things platforms enable remote patient monitoring, wearable sensing, telemedicine, personalized care, and smart rehabilitation, while raising issues of privacy, interoperability, implementation costs,

https://doi.org/10.3390/instruments10020030

and cybersecurity [13]. Objective measurements can improve diagnosis, treatment planning, patient follow-up, and rehabilitation outcome assessment, but clinical applications require high reliability, usability, and interpretability. Antonacci et al. focused on assessing shoulder muscle strength, comparing portable dynamometers with a reference load cell in a rigid serial configuration [14]. This example demonstrates that even seemingly simple clinical measurements require attention to reference systems, mechanical setup, signal processing, and concordance analysis. Wearable and rehabilitation-oriented instrumentation further confirms the relevance of integrated sensing: an instrumented ankle-foot orthosis integrating strain gauges, force-sensing elements, an inertial measurement unit, and Bluetooth Low Energy communication can provide objective information on the gait cycle for rehabilitation [15].

Agriculture and agri-food systems provide another example of metrology's expansion beyond traditional industrial applications. IoT, machine learning, and blockchain-based traceability can support crop monitoring, irrigation management, resource optimization, food safety, and fraud prevention [16,17]. Precision agriculture relies on distributed measurements of temperature, humidity, soil moisture, crop health, and environmental conditions. At the same time, agri-food supply chains require reliable data on production, storage, transportation, and quality [18]. The effectiveness of these digital solutions depends on the reliability and traceability of the underlying measurements.

In structural health monitoring, reproducibility and transferability are equally crucial. Chiominto et al. studied the reproducibility of impact area classification in aerospace structures using piezoelectric sensors, Lamb waves, signal processing, and machine learning models [19]. The study highlights that the performance of a classification procedure depends on feature extraction, acquisition conditions, training data, model selection, and impact point distribution. Similar issues arise in pipeline monitoring, where hydrophone-based acoustic sensing and machine learning have been investigated for robust leak detection and localization in the presence of variable sensor geometries and placements [20]. These examples confirm that machine learning-based measurement systems need to be validated not only for accuracy in a single experiment, but also for robustness, reproducibility, and transferability across devices, laboratories, and operating conditions.

Overall, the literature highlights a fundamental trend: measurement systems are becoming connected, intelligent, and application-oriented. The instrument is no longer an isolated device; it is part of a distributed architecture that includes sensing elements, integrated electronics, communication protocols, data storage, algorithms, user interfaces, and decision-making procedures. Consequently, measurement system design must consider not only sensitivity, resolution, and calibration, but also interoperability, latency, energy consumption, data security, model interpretability, and uncertainty propagation throughout the entire chain. Future research into instrumentation and measurement methods for Industry 4.0 and the IoT will therefore require a multidisciplinary approach. Sensors and acquisition systems must be designed in conjunction with data processing algorithms, physical models, validation procedures, and security mechanisms. Artificial intelligence must be combined with domain knowledge and metrological principles to obtain reliable and interpretable results. Finally, standardization, reproducible datasets, benchmarking procedures, and interlaboratory comparisons will be crucial for transferring innovative solutions from research prototypes to industrial, biomedical, agricultural, and smart environment applications. The future of Industry 4.0 and the IoT depends not only on the availability of data but also on the availability of reliable measurements. Data becomes useful only when its origin, quality, uncertainty, and meaning are known.

Acknowledgments: During the preparation of this manuscript, the author used ChatGPT V5.3 for the purposes of fixing all grammar, spelling, and punctuation errors and checking the standard of

https://doi.org/10.3390/instruments10020030

English. The author has reviewed and edited the output and takes full responsibility for the content of this publication.

Conflicts of Interest: The author declares no conflicts of interest.

References

1. Windmann, A.; Wittenberg, P.; Schieseck, M.; Niggemann, O. Artificial Intelligence in Industry 4.0: A Review of Integration Challenges for Industrial Systems. In *Proceedings of the 2024 IEEE 22nd International Conference on Industrial Informatics (INDIN)*; IEEE: New York, NY, USA, 2024. [CrossRef]
2. Nzumile, J.M.; Mahabi, V.; Taifa, I.W.R. Contribution of Industry 4.0 Technologies in Adopting Metrology 4.0 in Manufacturing Industries. In *Smart Engineering Management*; Management and Industrial Engineering; Springer: Cham, Switzerland, 2024; pp. 43–72. [CrossRef]
3. Rugbani, A. A Comprehensive Review of Smart Coordinate Measuring Machines: Architectural Innovations, Multi-Sensor Integration, and Uncertainty Evaluation for Industry 4.0. *Int. J. Precis. Eng. Manuf.* **2026**, *27*, 2247–2263. [CrossRef]
4. Boomgaarden, J.; Satterlee, N.; Schmitt, R.H.; Geiger, K.; Kang, J.S. Smart Metrology Techniques for Additive Manufacturing. *Int. J. Adv. Manuf. Technol.* **2026**, *144*, 261–289. [CrossRef]
5. Kombaya Touckia, J.V.; Henry, S. Towards Accurate Digital Twins: A Review of Calibration Methods Using Event-Based and Neuromorphic Vision. *Int. J. Adv. Manuf. Technol.* **2026**. [CrossRef]
6. Bitencourt, J.; Wooley, A.; Harris, G. Verification and Validation of Digital Twins: A Systematic Literature Review for Manufacturing Applications. *Int. J. Prod. Res.* **2025**, *63*, 342–370. [CrossRef]
7. Cação, J.; Santos, J.; Antunes, M. Explainable AI for industrial fault diagnosis: A systematic review. *J. Ind. Inf. Integr.* **2025**, *47*, 100905. [CrossRef]
8. Martiri, L.; Esmaili, P.; Moschetti, A.; Cristaldi, L. Deep-Learning-Based Bearing Fault Classification Using Vibration Signals Under Variable-Speed Conditions. *Instruments* **2025**, *9*, 33. [CrossRef]
9. Somu, N.; Dasappa, N.S. An Edge-Cloud IIoT Framework for Predictive Maintenance in Manufacturing Systems. *Adv. Eng. Inform.* **2025**, *65*, 103388. [CrossRef]
10. Jain, P.; Mondal, K.K.; Das, D.; Khandelwal, A. Intelligent Predictive Maintenance: Multivariate ML Model Optimization in an Edge-Fog-Cloud Environment. *Computing* **2025**, *107*, 225. [CrossRef]
11. Khan, F.M.; Zeb, A.; Rahman, T.; Al-Khasawneh, M.A.; Daradkeh, Y.I.; Siddiqui, I.F.; Bashir, A.K.; Ullah, I. XAI-Driven Data Mining for Self-Defending IoT Systems: Enhancing Cybersecurity Transparency in the Age of Smart Cities. *Cogn. Comput.* **2026**, *18*, 16. [CrossRef]
12. Guo, X.; Fei, T.; Wang, Z.; Xu, X.; Mukhopadhyay, S.; Bhardwaj, A.; Jin, Z. A Review of Next-Generation Sensor Technologies for Human-Machine Interaction. *IEEE Sens. J.* **2025**, *26*, 14252–14269. [CrossRef]
13. Shah, T.; Dumasia, K.; Lad, H.; Shah, K.; Nadiyadra, K.; Suratwala, V. A Systematic Review of Emerging Trends of IoT in Healthcare and IoMT Frameworks. *Discov. Public Health* **2026**, *23*, 52. [CrossRef]
14. Antonacci, C.; Carnevale, A.; Mancini, L.; de Sire, A.; D'Hooghe, P.; Mercurio, M.; Papalia, R.; Schena, E.; Longo, U.G. Shoulder Muscle Strength Assessment: A Comparative Study of Hand-Held Dynamometers and Load Cell Measurements. *Instruments* **2026**, *10*, 2. [CrossRef]
15. Mahraoui, S.; Serpelloni, M. Wireless Instrumented Ankle Foot Orthosis (AFO) for Gait Cycle Monitoring. *Instruments* **2026**, *10*, 23. [CrossRef]
16. Sudha, S.P.; Loret, J.B.S. A Review on Machine Learning-Based Precision Agriculture Techniques for Crop Farming Monitoring with IoT. *Discov. Environ.* **2026**, *4*, 10. [CrossRef]
17. Khoualdi, J.; Abdelhedi Abdelmoula, I.; Roth, U.; Kaffel Ben Ayed, H.; Badis, H. Blockchain- and IoT-Based Agri-Food Supply Chain Traceability to Improve Food Safety and Combat Food Waste and Food Fraud. *IEEE Access* **2026**, *14*, 59624–59658. [CrossRef]
18. Liu, Z.; Yu, X.; Liu, N.; Liu, C.; Jiang, A.; Chen, L. Integrating AI with detection methods, IoT, and blockchain to achieve food authenticity and traceability from farm-to-table. *Trends Food Sci. Technol.* **2025**, *158*, 104925. [CrossRef]
19. Chiominto, L.; D'Emilia, G.; Gaspari, A.; Natale, E.; Nicassio, F.; Scarselli, G. A Procedure for Performing Reproducibility Assessment of the Accuracy of Impact Area Classification for Structural Health Monitoring in Aerospace Structures. *Instruments* **2026**, *10*, 6. [CrossRef]
20. Zuo, X.; Satterlee, N.; Guwalani, R.; Park, C.-W.; Kang, J.S. Comparative Analysis of Machine Learning Approaches for Leak Detection in Pipelines Using Hydrophone Acoustic Sensing. *Measurement* **2026**, *261*, 120005. [CrossRef]

https://doi.org/10.3390/instruments10020030

instruments

MDPI

Communication

Conceptual Study on the Implementation of NRTA for Industrial Applications

Melissa Azzoune [1,*], Ludovic Mathieu [1,*], Ngoc Duy Trinh [2,3], Mourad Aïche [1], Laurence Villatte [2], Fabrice Piquemal [1], Lionel Tondut [3] and Sylvain Pelletier [2]

1 LP2i UMR5797, Université de Bordeaux, CNRS, 33170 Gradignan, France
2 Orano Recyclage, 125 Avenue de Paris, 92320 Châtillon, France
3 Orano Recyclage, Établissement de la Hague, 50444 La Hague, France
* Correspondence: azzoune@lp2ib.in2p3.fr (M.A.); mathieu@lp2ib.in2p3.fr (L.M.)

Abstract: Neutron Resonance Transmission Analysis (NRTA) is a non-destructive technique allowing the elemental and isotopic characterization of materials and objects. This study represents a first step toward understanding the NRTA technique and developing a novel compact system adapted for industrial applications. The industrial feasibility of the NRTA was assessed by simulating a compact system using the Monte Carlo code MCNP 6.1. Neutron transmission spectra were generated for various metallic samples, ranging from 0.1 mm to 1 cm in thickness, and analyzed using a home-developed quantification method that incorporates nuclear cross sections from the ENDF/B-VIII.0 library and accounts for instrumental resolution. For this first study, an idealized configuration was considered, with a 0 μs pulsed neutron source and a Gaussian resolution function, to validate the methodology under a simple controlled condition. The results demonstrate that the areal densities of isotopes of Uranium and Plutonium can be determined with relative deviations below 10%, even under compact measurement conditions. This study validates the characterization method and represents a first step toward the continued development of an industrial NRTA prototype for rapid, non-destructive isotopic control of nuclear materials.

Keywords: neutron resonance transmission analysis; time-of-flight; transmission measurement; resonance

1. Introduction

Since the mid-1970s, France has strategically chosen to reprocess and recycle spent nuclear fuel. This policy aimed to optimize the use of uranium resources by reusing separated uranium and plutonium, while reducing the volume and radiotoxicity of the final waste [1]. The high-level waste generated by these operations is vitrified and stored at the La Hague Orano facility [2], which today serves as the international industrial reference center for such processes. In 2024, the French government reaffirmed the continuation of this policy beyond 2040, planning the extension of existing facilities and the study of the construction of new infrastructures [3]. In this context, precise isotopic analyses remain a major need to ensure the quality, safety, and traceability of reprocessing operations.

Within the nuclear fuel cycle, a precise knowledge of the isotopic composition of fissile and fertile materials, in particular uranium, plutonium, and minor actinides in different steps of the recycling process, is essential to optimize reprocessing procedures and ensure the safety of the facilities. To date, inductively coupled plasma mass spectrometry (ICP-MS) [4] remains the reference technique due to its high isotopic precision. However, this method is destructive, complex to implement, and requires several days of chemical preparation and analysis. Non-destructive reference nuclear measurement techniques, such as neutron or gamma spectroscopy (in active and passive modes), have also been employed for decades [5] with high quality. Nevertheless, these nuclear measurement techniques lack the capacity to characterize all present isotopes (e.g., ^{242}Pu) and depend on reactor calculation codes for data interpretation.

The development of a novel, rapid, non-destructive method capable of providing reliable isotopic compositions while being less dependent on experimental conditions would therefore represent a major advance for some specific nuclear characterization applications. In this context, Neutron Resonance Transmission Analysis (NRTA) appears as a promising alternative. Due to its isotopic selectivity and its ability to cover a wide range of nuclei (medium- and high-Z isotopes) [6], this technique offers a unique potential to improve the characterization of nuclear materials under industrial conditions.

Historically, NRTA has been implemented since the 1970s in research centers to determine the isotopic composition of fresh and irradiated fuels [7], and especially for total neutron reaction cross-section measurements [8]. Nevertheless, its use has remained limited to fundamental research centers (for studying thin and homogeneous samples) due to experimental constraints. Indeed, this technique requires long flight paths (from tens up to few hundred meters, as in GELINA—JRC-Geel [9]) and high-performance accelerators capable of producing intense ($\approx 10^{13}$ n/s [9]), high-resolution neutron beams (lower than μs [9]), necessary, as well as the use of sophisticated data analysis tools such as REFIT [10], SAMMY [11], or CONRAD [12].

The current challenge is to design a more compact system, suitable for analyzing thick and non-homogeneous samples, capable of delivering results within minutes or hours, with an expected accuracy on the order of a few percent. Several international studies are exploring this approach. In the United States, for instance, prototypes using DT neutron generators and compact detectors have demonstrated the feasibility of isotopic identification of uranium and plutonium [13,14]. In Japan, the Japan Atomic Energy Agency is developing a compact NRTA system for existing facilities [15,16], including simulation studies and pulsed neutron generation [17,18]. In France, the Orano—CNRS Bordeaux (LP2i) partnership is part of this effort, aiming to develop an industrial NRTA prototype dedicated to spent nuclear fuel reprocessing needs.

This study investigates the industrial feasibility of NRTA, focusing on its adaptation to compact systems under realistic industrial constraints: very short flight paths, neutron sources with limited flux and time resolution, and complex thick sample configurations. The objective of this work is to present the first results of the developed NRTA characterization method. The method was designed and developed using MCNP [19] simulations rather than experimental data. This preliminary stage aims to test and validate the methodology under idealized conditions prior to its future experimental application. Results from Monte Carlo simulations are presented, along with a quantification method developed for analyzing neutron resonance transmission, paving the way for the practical application of this technique in an industrial context.

2. Neutron Resonance Transmission Analysis (NRTA)

2.1. Principle and Experimental Conditions

NRTA is a non-destructive analytical technique that enables the identification and quantification of the elemental and isotopic composition of materials based on their characteristic total cross-section resonances in the epithermal energy range.

This active neutron interrogation method relies on the interaction of a pulsed neutron beam, moderated to thermal and epithermal energies, with a sample. These energy ranges correspond to the regions where resonances appear in the neutron-induced reaction cross sections.

As the neutrons pass through the sample, some are absorbed or scattered, while the remaining fraction is measured by a detector placed downstream, using the time-of-flight (TOF) technique. The fraction of neutrons that pass through the sample without interaction constitutes the main quantity measured in NRTA.

In the ideal case of a parallel beam perpendicular to a homogeneous plate, this fraction, called transmission T, is expressed as:

$$T_{Theo}(E) = \mathbf{exp}\left[-\sum_k n_k \sigma_{tot,k}(E)\right] \quad (1)$$

where $\sigma_{tot,\,k}$ (E) represents the Doppler broadened total cross section of isotope k, and n_k its areal density.

Experimentally, the transmission is derived from the ratio of flux measured with and without the sample:

$$T_{exp}(t) = \frac{\varphi_{in} - B_{in}}{\varphi_{out} - B_{out}} \quad (2)$$

where

φ_{in} is the flux measured with the sample in the beam,
φ_{out} is the flux measured without the sample,
B_{in} and B_{out} are the corresponding background counts for the *in* and *out* configurations.

2.2. Resolution Function

In NRTA experiments, measurements provide only the transmission spectra as a function of time using the TOF technique. To obtain the energy dependence, the measured time-of-flight must be converted into the neutron kinetic energy using the non-relativistic relation.

However, experimental data are never perfectly resolved. As neutrons travel through various materials (moderator, target, detector, etc.), their flight times are affected by multiple effects such as moderation time, burst width of the accelerator, and detector response. Consequently, neutrons with the same kinetic energy do not necessarily arrive at the detector simultaneously.

To better understand this phenomenon, a simulation was performed using the geometry illustrated in Figure 1. The setup includes:

1. Pulsed source: 14 MeV neutrons produced by a pulsed neutron generator (D–T), with a 0 µs duration, positioned at the center of the moderator.
2. Moderator block: slowing down neutrons to reach the resonance energies of the target isotopes (<100 eV, epithermal range). The moderator is modeled as a 20 × 20 × 20 cm^3 polyethylene block.
3. Collimation tube: serving as a perfect neutron guide by removing all off-axis neutrons.
4. Sample: where thermalized neutrons interact with the sample nuclei, producing absorption signatures characteristic of each isotope. The sample is modeled as a metallic disc of 14 cm in diameter and variable thickness, ranging from 0.1 mm to 1 cm depending on the studied case.
5. Detector: where neutron energy is detected using the time-of-flight method ($E_n = \frac{1}{2} m_n \left(\frac{d_{TOF}}{t_{TOF}}\right)^2$). In the simulations, the detector is modeled as an ideal circular disk with the same diameter as the sample, ensuring perfect collection of all transmitted neutrons without any efficiency loss.
6. Air (shown in grey on Figure 1): the surrounding area where no neutrons are simulated.

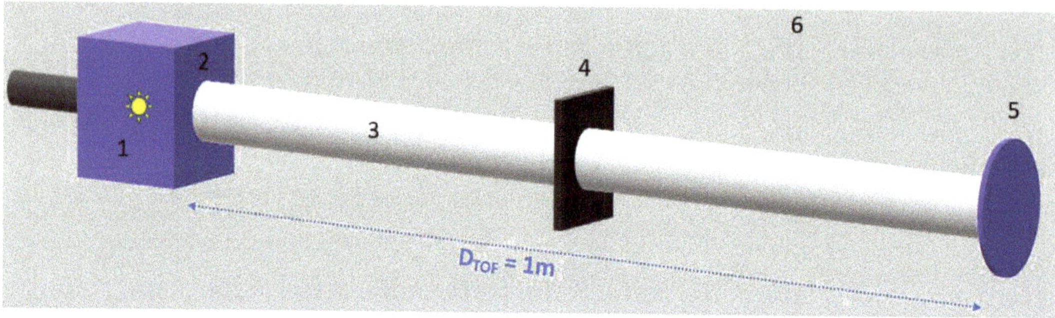

Figure 1. Compact NRTA system geometry used in MCNP simulations, illustrating (1) the neutron source, (2) polythene moderator, (3) collimation tube, (4) sample, (5) detector, and (6) air.

For this simulation, and for all subsequent results presented in this work, a total of NPS = 10^{11} particles were used, leading to a computation time of approximately 8 h, with variance reduction techniques applied to optimize the calculation. All simulations were performed under idealized conditions: the neutron source had zero pulse width, the detector response was ideal, and the collimation tube acted as a perfect neutron guide (removing all off-axis neutrons).

The results were collected at a 1 m flight path using F1 tallies (current over the surface): one tally recorded the neutron time-of-flight, while another tally collected the corresponding neutron energies. Figure 2 clearly illustrates the degradation of energy information caused by the system's time resolution by comparing the simulated energy spectrum (black) with the corresponding time-converted spectrum (red). Unlike experimental measurements, Monte Carlo simulations provide direct access to the true energy spectrum, allowing a clear visualization of how time-resolution effects broaden and distort the resonances under experimental conditions.

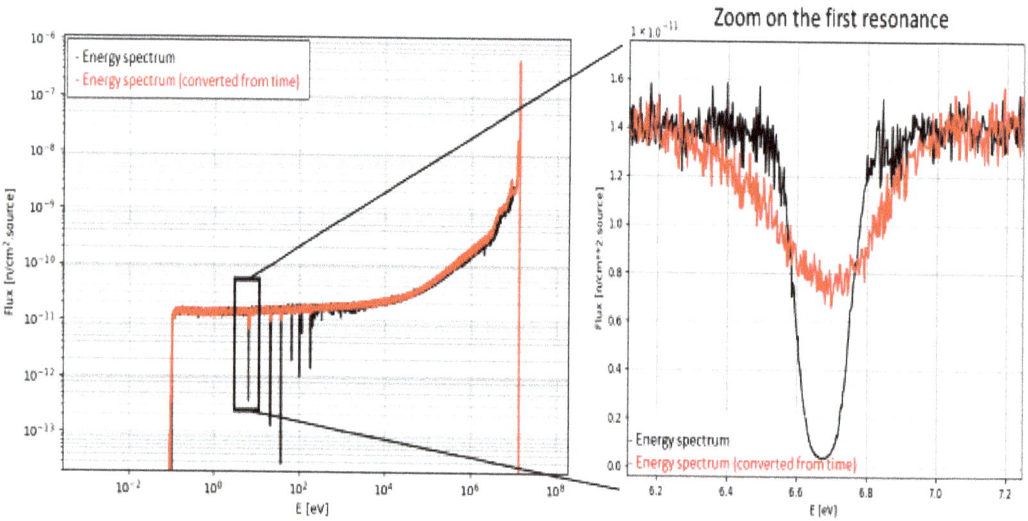

Figure 2. Left: Simulated energy spectrum (black) vs. time-converted spectrum (red). **Right**: Zoom on the first resonance (^{238}U sample).

As a result, direct isotopic characterization from the areas of resonance dips cannot be performed:

- Resonances appear broadened and shallower, making auto-absorption corrections impossible;
- Resonances from other isotopes may overlap significantly.

Consequently, an accurate characterization of the samples requires accounting for these different system contributions in order to correct the theoretical spectra. This correction is achieved through the so-called *resolution function*, which describes the overall time- and energy-broadening effects of the system. In a realistic NRTA system, the system's resolution function mainly results from two contributions:

1. The source pulse width, which introduces temporal spreading (not simulated here);
2. The moderating process, which scatters the neutron arrival times for a given energy.

By accounting for the system's resolution function in the calculation of the theoretical transmission, it becomes possible to retrieve the quantities of isotopes present in the sample. Indeed, the corrected theoretical transmission can be directly related to the experimental transmission T_{exp} provided that the basic assumptions of transmission experiments are fulfilled [20]:

- The sample is perpendicular to a parallel incoming neutron beam.
- All detected neutrons have passed through the sample.
- Neutrons scattered by the sample are not registered by the neutron detector.

3. Methodology

As discussed previously, the system's resolution limits the ability to directly characterize the isotopic composition from resonance dips. To overcome these limitations and fully exploit the information contained in the transmission spectra, a specific analysis method has been developed. This method is based on the use of a Python 3 script and is designed not only to process theoretical spectra together with simulated or experimental results, but also to provide a user-friendly code adapted for industrial applications, unlike existing, more complex and sophisticated analysis tools such as REFIT [10], SAMMY [11], or CONRAD [12]. The adopted approach is structured around three main steps, which are also summarized in Figure 3:

1. Generation of theoretical transmission spectra: The theoretical transmission spectra were calculated using microscopic neutron total cross sections from the ENDF/B-VIII.0 library at 293.6 K, according to the Beer–Lambert law (Equation (1)), based on the assumed isotopic composition and areal densities of the sample. For the calculation, initial bounds are set for each material quantity (n), which are then iteratively refined using the differential evolution algorithm (implemented in SciPy library [21]) to fit the theoretical spectrum to the simulated (or experimental data).
2. Spectral degradation using a resolution function: the theoretical transmission spectrum is convoluted with a resolution function representing the characteristics of the NRTA system. This function accounts in particular for the moderation process and the flight path length. The goal is to obtain a theoretical spectrum comparable to the simulated or experimental spectrum.
3. Comparison of theory vs. data (from Monte Carlo simulation or experiment): the data spectrum is compared in each energy bin to the theoretical spectrum. A global minimum is determined using the differential evolution algorithm, which iteratively adjusts the material quantities (n) to achieve the best agreement between theory and simulation, as illustrated in Figure 3.

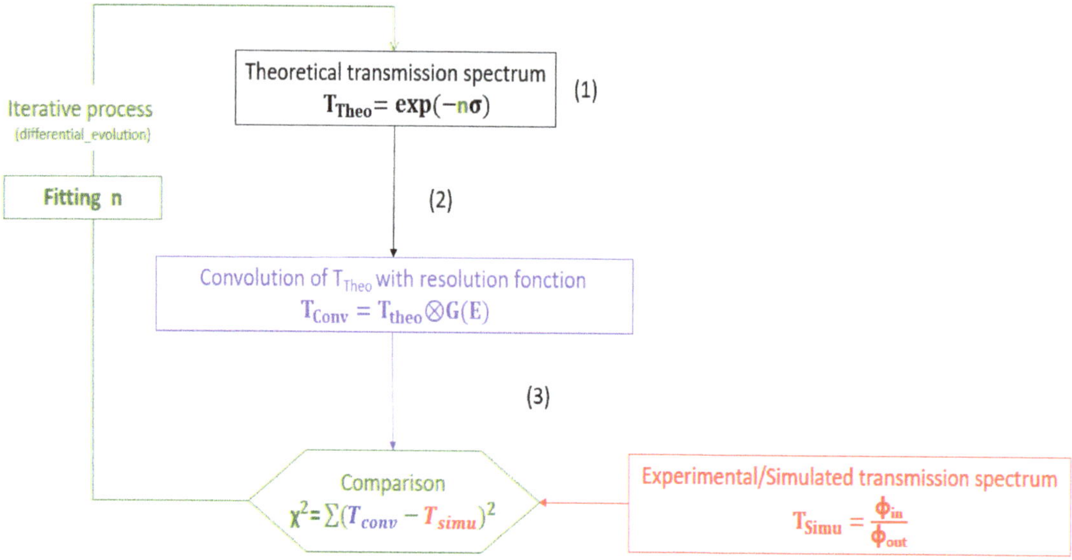

Figure 3. Schematic of the NRTA quantification method. The main steps include (1) generation of theoretical transmission spectra from known nuclear cross sections, (2) transmission degradation using the resolution function representing the compact system, and (3) comparison between theoretical and simulated/experimental spectra to adjust the isotopic areal densities via a least-squares fit.

4. Results

To assess the validity of the characterization method, simulations were performed on metallic disc samples with thicknesses ranging from 0.1 mm to 1 cm. The simulated setup corresponds to the compact configuration presented in Figure 1, featuring a simplified transmission geometry with an ideal neutron source, an ideal detector, and the condition that all detected neutrons have passed through the sample.

Figure 4 illustrates an example of a simulated neutron transmission spectrum obtained for a 1 cm thick metallic ^{238}U sample. The resonance structures characteristic of uranium are clearly visible, confirming that the configuration allows the observation of individual resonances even under compact conditions.

The simulated transmission spectra were then analyzed using the developed characterization method (Figure 3), which relies on nuclear cross-sections from the ENDF/B-VIII.0 data library. For this first study, and to obtain preliminary results, a Gaussian resolution function was chosen as a simplified approach to represent the instrumental response of the compact system. Since the Gaussian approximation closely represents the resolution function of a system with a negligible pulse width, an ideal neutron source with zero pulse width was considered to validate the method before extending it to more realistic configurations. The standard deviation of this Gaussian varies with energy to account for the time-of-flight dependence of the system resolution.

A quantitative comparison between the simulated areal densities and those obtained using the NRTA quantification method is reported in Table 1. For thin mono-isotopic samples, the relative deviations between simulated and retrieved values remain below 5%. In the case of mixed or thicker samples, the deviations remain below 10%.

Overall, the developed NRTA method predicts isotopic areal densities with a relative accuracy generally between 5% and 10% and the observed discrepancies mainly arise from the simplified choice of a Gaussian resolution function. In practice, the actual instrumental response, extracted from Monte Carlo simulations, exhibits an asymmetric shape. Incorporating this more realistic profile into the analysis is expected to further reduce deviations and improve the predictive accuracy of the method. It should be noted that, to date, no sensitivity or uncertainty study has been conducted on the present simulation results; performing such analyses in the future would allow a more thorough assessment of the deviations between simulated areal densities and those obtained using the NRTA quantification method.

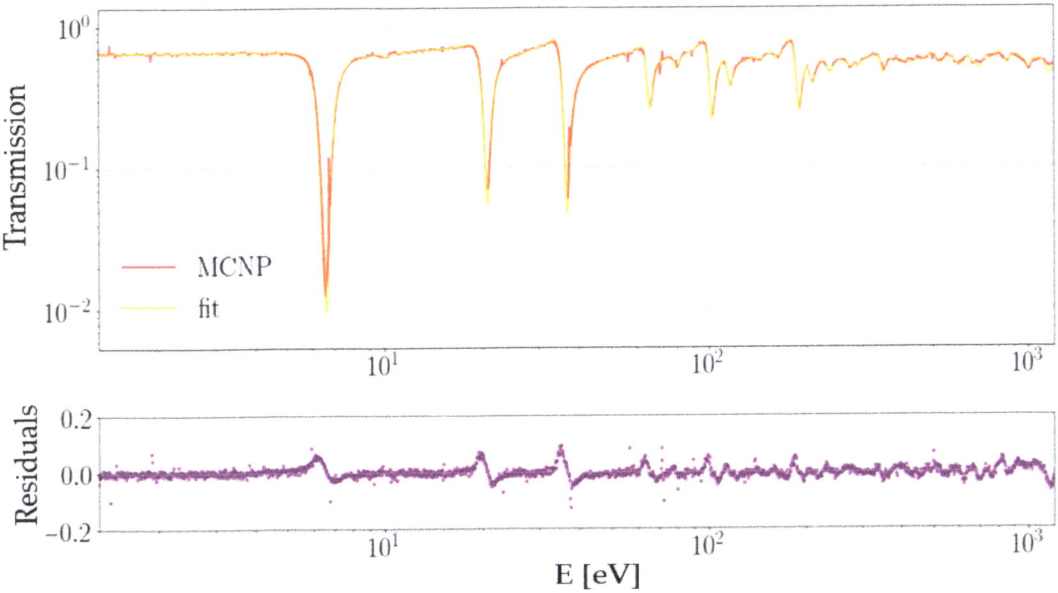

Figure 4. Transmission through a 1 cm-thick ^{238}U sample, as detailed in Table 1. The simulated transmission is shown in red, the fitted result in yellow, and the difference between the two is displayed as a residual plot in purple at the bottom of the figure.

Table 1. Results of the NRTA. The simulations were performed for a 1 m flight path and an ideal source with a pulse width of 0 µs.

Samples		Thickness	Areal Number Density (at/b) Simulated (S)	Areal Number Density (at/b) Analyses (M)	Ratio (S/M)
^{238}U		0.1 mm	4.830×10^{-4}	4.637×10^{-4}	1.042
^{238}U		1 cm	4.830×10^{-2}	4.492×10^{-2}	1.075
^{238}U	50%	0.1 mm	2.415×10^{-4}	2.222×10^{-4}	1.087
^{238}Pu	50%			2.222×10^{-4}	1.087
^{238}U	50%	1 mm	2.415×10^{-3}	2.502×10^{-3}	0.965
^{235}U	50%			2.391×10^{-3}	1.01

5. Conclusions

This study constitutes a first step toward both understanding the NRTA technique and familiarizing ourselves with its practical implementation through simulations. The initial approach focused on idealized conditions, using a 0 µs pulsed source and a Gaussian resolution function, to validate a characterization methodology specifically developed for industrial applications. Under these ideal conditions, the method appears capable of estimating the isotopic compositions of metallic samples, reproducing areal densities with relative deviations generally below 10%, even in compact system configurations.

This conceptual study establishes a reference framework for the industrial feasibility of the Neutron Resonance Transmission Analysis (NRTA) technique and demonstrates the robustness of the proposed approach. Future developments will include refining the characterization method by incorporating a more realistic, asymmetric resolution function derived from simulations and accounting for the intrinsic time resolution of the neutron source. The methodology will also be extended to handle more complex isotopic mixtures, and a dedicated uncertainty and sensitivity study will be performed. These improvements will allow us to approach a more realistic NRTA system and provide a solid basis to evaluate and compare our results with those obtained by other research groups.

In parallel, experimental validation campaigns are planned at GELINA and MONNET facilities, in collaboration with JRC-Geel. These measurements will play a key role in confirming the applicability of the proposed approach and represent a significant step toward the industrial deployment of NRTA for rapid and non-destructive isotopic control of nuclear materials.

Author Contributions: Conceptualization, M.A. (Melissa Azzoune) and L.M.; methodology, M.A. (Melissa Azzoune) and L.M.; software, M.A. (Melissa Azzoune); validation, L.M. and N.D.T.; formal analysis, M.A. (Melissa Azzoune); writing—original draft preparation, M.A. (Melissa Azzoune); writing—review and editing, M.A. (Melissa Azzoune), L.M. and N.D.T.; supervision, L.M., N.D.T., M.A. (Mourad Aïche), F.P., L.V., L.T. and S.P.; project administration, L.V.; funding acquisition, L.V. All authors have read and agreed to the published version of the manuscript.

Funding: The authors declare that this study received funding from Orano company.

Data Availability Statement: The original contributions presented in this study are included in the article. Further inquiries can be directed to the corresponding author(s).

Conflicts of Interest: Authors L.M., M.A. (Mourad Aïche) and F.P. collaborate with the Orano company on other projects. Authors N.D.T., L.V., L.T. and S.P. are employed by the Orano company. Author M.A. (Melissa Azzoune) is funded by the Orano company.

References

1. OCDE; Agence Pour L'Énergie Nucléaire. *Cycles du Combustible Nucléaire Avancés et Gestion des Déchets Radioactifs*; Développement de L'Énergie Nucléaire; OECD: Paris, France, 2006; ISBN 978-92-64-02487-8.
2. Orano Recyclage. *Rapport d'Information du Site Orano la Hague*; Orano Recyclage: Paris, France, 2023.
3. Benjamin, M. Focus: France seeks strategy as nuclear waste site risks saturation point. *Reuters*. 3 February 2023. Available online: https://www.reuters.com/business/environment/france-seeks-strategy-nuclear-waste-site-risks-saturation-point-2023-02-03 (accessed on 30 October 2025).
4. Krachler, M.; Alvarez-Sarandes, R.; Van Winckel, S. Challenges in the quality assurance of elemental and isotopic analyses in the nuclear domain benefitting from high resolution ICP-OES and sector field ICP-MS. *J. Radioanal. Nucl. Chem.* **2015**, *304*, 1201–1209. [CrossRef] [PubMed]
5. Benefits and Drawbacks of Non-Destructive Assay (NDA) for Nuclear Safeguards—An Overview—European Commission. Available online: https://esarda.jrc.ec.europa.eu/publications/benefits-and-drawbacks-non-destructive-assay-nda-nuclear-safeguards-overview_en (accessed on 30 October 2025).
6. Paradela, C.; Alaerts, G.; Becker, B.; Heyse, J.; Kopecky, S.; Moens, A.; Mondelaers, W.; Schillebeeckx, P.; Wynants, R.; Harada, H.; et al. Characterization of nuclear material by Neutron Resonance Transmission Analysis. *Nuovo C. C Geophys. Space Phys. C* **2016**, *38*, 176. [CrossRef]
7. Behrens, J.W.; Johnson, R.G.; Schrack, R.A. Neutron Resonance Transmission Analysis of Reactor Fuel Samples. *Nucl. Technol.* **1984**, *67*, 162–168. [CrossRef]
8. Schillebeeckx, P.; Becker, B.; Danon, Y.; Guber, K.; Harada, H.; Heyse, J.; Junghans, A.R.; Kopecky, S.; Massimi, C.; Moxon, M.C.; et al. Determination of Resonance Parameters and their Covariances from Neutron Induced Reaction Cross Section Data. *Nucl. Data Sheets* **2012**, *113*, 3054–3100. [CrossRef]

9. Schillebeeckx, P.; Borella, A.; Emiliani, F.; Gorini, G.; Kockelmann, W.; Kopecky, S.; Lampoudis, C.; Moxon, M.; Cippo, E.P.; Postma, H.; et al. Neutron resonance spectroscopy for the characterization of materials and objects. *J. Instrum.* **2012**, *7*, C03009. [CrossRef]
10. Moxon, M.C.; Ware, T.C.; Dean, C.J. *REFIT-2009: Multilevel Resonance Parameter Least Squares Fit of N Transmission, Capture, Fission & Self Indication Data*; NEA-0914; OECD Nuclear Energy Agency: Paris, France, 2009.
11. Larson, N.M. *Updated User's Guide for Sammy: Multilevel R-Matrix Fits to Neutron Data Using Bayes' Equations*; Oak Ridge National Laboratory (ORNL): Oak Ridge, TN, USA, 2008.
12. Archier, P.; Jean, C.D.S.; Kopecky, S.; Litaize, O.; Noguère, G.; Schillebeeckx, P.; Volev, K. Recent Developments in the CONRAD Code regarding Experimental Corrections. *EPJ Web Conf.* **2013**, *42*, 02004. [CrossRef]
13. Klein, E.A. Neutron Resonance Transmission Analysis of Nuclear Material Using a Portable D-T Neutron Generator. Ph.D. Thesis, Massachusetts Institute of Technology, Cambridge, MA, USA, 2023.
14. McDonald, B.S.; Danagoulian, A.; Gilbert, A.J.; Klein, E.A.; Kulisek, J.A.; Moore, M.E.; Rahon, J.M.; Zalavadia, M.A. Neutron resonance transmission analysis prototype system for thorium fuel cycle safeguards. *Nucl. Instrum. Methods Phys. Res. Sect. Accel. Spectrometers Detect. Assoc. Equip.* **2024**, *1062*, 169148. [CrossRef]
15. Tsuchiya, H.; Kitatani, F.; Maeda, M.; Toh, Y.; Kureta, M. Development of Neutron Resonance Transmission Analysis as a Non-Destructive Assay Technique for Nuclear Nonproliferation. *Plasma Fusion Res.* **2018**, *13*, 2406004. Available online: https://www.jstage.jst.go.jp/article/pfr/13/0/13_2406004/_article/-char/en (accessed on 30 October 2025). [CrossRef]
16. Guembou Shouop, C.J.; Tshuchiya, H. Monte Carlo and experimental assessment of the optimal geometry of the source and collimator for a table-top NRTA system for small nuclear material measurement. *Nucl. Instrum. Methods Phys. Res. Sect. Accel. Spectrometers Detect. Assoc. Equip.* **2025**, *1072*, 170189. [CrossRef]
17. E-Journal of Advanced Maintenance (EJAM). Available online: https://www.jsm.or.jp/ejam/Vol.11No.1/AA/AA156/AA156.html (accessed on 30 October 2025).
18. Toh, Y.; Ohzu, A.; Tsuchiya, H.; Furutaka, K.; Kitatani, F.; Komeda, M.; Maeda, M.; Koizumi, M. Development of active neutron nondestructive assay system. *EPJ Web Conf.* **2025**, *329*, 06001. [CrossRef]
19. Pelowitz, D.B.; Goorley, J.T.; James, M.R. *MCNP6 User's Manual*; Technical Report LA-CP-13-00634; Los Alamos National Laboratory: Los Alamos, NM, USA, 2013.
20. Paradela, C.; Wynants, R.; Kopecky, S.; Salamon, L.; Kureta, M.; Alaerts, G.; Harada, H.; Tsuchiya, H.; Koizumi, M. *Neutron Resonance Analysis: System Requirements*; Publications Office of the European Union: Luxembourg, 2016; ISBN 978-92-79-63918-0.
21. Storn, R.; Price, K. Differential Evolution—A Simple and Efficient Heuristic for global Optimization over Continuous Spaces. *J. Glob. Optim.* **1997**, *11*, 341–359. [CrossRef]

instruments

MDPI

Article

Deep-Learning-Based Bearing Fault Classification Using Vibration Signals Under Variable-Speed Conditions †

Luca Martiri *, Parisa Esmaili, Andrea Moschetti and Loredana Cristaldi

Dipartimento di Elettronica, Informazione e Bioingegneria, Politecnico di Milano, 20133 Milan, Italy; loredana.cristaldi@polimi.it (L.C.)

* Correspondence: luca.martiri@polimi.it

† This paper is an extended version of our paper published in IEEE MetroInd4.0 & IoT 2025, Castelldefels, Spain, 1–3 July 2025.

Abstract: Predictive maintenance in industrial machinery relies on the timely detection of component faults to prevent costly downtime. Rolling bearings, being critical elements, are particularly prone to defects such as outer race faults and ball spin defects, which manifest as characteristic vibration patterns. In this study, we introduce a novel bearing vibration dataset collected on a testbench under both constant and variable rotational speeds (0–5000 rpm), encompassing healthy and faulty conditions. The dataset was used for failure classification and further enriched through feature engineering, resulting in input features that include raw acceleration, signal envelopes, and time- and frequency-domain statistical descriptors, which capture fault-specific signatures. To quantify prediction uncertainty, two different approaches are applied, providing confidence measures alongside model outputs. Our results demonstrate the progressive improvement of classification accuracy from 87.2% using only raw acceleration data to 99.3% with a CNN-BiLSTM (Convolutional Neural Network–Bidirectional Long Short-Term Memory) ensemble and advanced features. Shapley Additive Explanation (SHAP)-based explainability further validates the relevance of frequency-domain features for distinguishing fault types. The proposed methodology offers a robust and interpretable framework for industrial fault diagnosis, capable of handling both stationary and non-stationary operating conditions.

Keywords: fault diagnosis; rolling bearings; predictive maintenance; neural networks; uncertainty quantification

1. Introduction

Modern industrial machinery is increasingly complex, necessitating proactive strategies to ensure reliability and minimize downtime. Predictive maintenance has emerged as a key solution that exploits sensor data and Machine Learning (ML) models, enabling early detection of potential faults. Consequently, it reduces unplanned downtime and optimizes maintenance schedules, extending the operational lifespan of equipment while reducing costs [1].

Recent advances in ML have significantly improved Fault Diagnosis and Prognosis (FDP) [2]. By analyzing large volumes of sensor data, ML algorithms can detect patterns and anomalies indicative of component failures. This allows for both real-time detection and predictive insights, supporting optimized maintenance schedules and improved overall machine reliability.

Among critical components, rolling bearings are particularly prone to faults such as wear, misalignment, and ball defects [3]. Failure of these elements can propagate severe damage throughout machinery, making real-time monitoring essential [4]. Vibration signals, temperature fluctuations, and other performance indicators provide rich information for assessing bearing health, detecting early-stage faults, and informing timely maintenance decisions.

This work is an extension of the paper "A Novel Benchmark for Fault Detection in Rolling Bearings Using CNNs and Monte Carlo Dropout" [5]. Building upon the dataset introduced in the conference paper, we develop a classification framework based on different neural network architectures with white-box feature selection informed by domain knowledge to operate both in constant and variable speed conditions. Features include raw accelerations, envelopes, and statistical and spectral descriptors. To increase trust in predictions, Monte Carlo (MC) dropout is applied for uncertainty quantification.

The remainder of this paper is structured as follows: Section 2 describes the problem and the challenges of bearing fault detection. Section 3 details the experimental setup, dataset creation, feature extraction, and model architectures. Section 4 presents classification results, model comparisons, and explainability analyses. Finally, Section 5 concludes the study and outlines directions for future research.

2. Problem Description

Bearings are critical components in many industrial machines, experiencing wear and tear throughout their operational life. Over time, this wear can result in increased vibrations, causing different faults, and finally leading to failures and costly downtime [6]. To prevent these scenarios, the health of the bearings must be continuously monitored using sensors, i.e., accelerometers, in order to analyze real-time acceleration data. This, combined with a proactive maintenance approach, could lead to early identification of failures, allowing for programmed interventions to prevent more severe damage.

2.1. Physical Mechanisms of Rolling Bearing Faults

Rolling element bearings are composed of an inner race, an outer race, rolling elements (balls or rollers), and a cage. Localized defects typically initiate as pits or spalls on one of these elements and, as each rolling element passes over the damaged zone [7]. These defects produce distinct vibration signatures, which are determined by the kinematics of the rolling elements and the spatial relationship between the defect and the load zone. For faults located on the outer race, the damage remains stationary relative to the radial load, resulting in each rolling element generating an impact of similar amplitude as it traverses the defect. This process produces a nearly periodic train of impulses at the Ball Pass Frequency Outer Race (BPFO), with its harmonic series clearly observable in both the vibration envelope and the demodulated spectrum. In contrast, defects on rolling elements generate a different signature. Here, the defect rotates with the ball and enters the load region only intermittently, leading to amplitude-modulated bursts centered at the Ball Spin Frequency (BSF) with characteristic sidebands at the Fundamental Train Frequency (FTF). Inner-race defects display hybrid characteristics, producing impulses at the Ball Pass Frequency Inner Race (BPFI) as the rotating defect aligns with the load direction once per shaft revolution. These defects also induce shaft-speed sidebands due to periodic variations in the dynamic transfer path [8,9]. These classical analytical bearing models, when combined with modern cyclostationary-based interpretations, account for periodic impulses, modulation sidebands, and resonance-rich transients in measured vibration signals [7,10,11]. Consequently, such physics-based understanding underpins the diagnostic interpretation and feature design utilized in the present study.

2.2. Signal Characteristics and Feature Foundations for Fault Diagnosis

Low computing burden and straightforward interpretability make time- and frequency-domain indicators essential for rolling–bearing condition monitoring. The Root Mean Square (RMS) metric is widely used as an energy-based severity indicator for time, which becomes more excited by fault-induced impacts. This behavior is consistently studied in both industrial and controlled degradation studies. Higher-order statistics, such as kurtosis, skewness, and crest factors, have also demonstrated strong effectiveness in identifying early-stage localized defects [12]. However, in multiscale fault-severity analyses, alternative metrics such as the Energy Index (EI), which compare segmental and global RMS levels, have been suggested due to the known non-monotonicity of kurtosis at advanced deterioration stages [13]. Frequency-domain analysis, however, provides a direct mapping between bearing kinematics and spectral content, enabling the identification of characteristic fault frequencies together with their modulation-induced sidebands. Spectral decomposition, which is performed using band-pass filtering, Hilbert transform demodulation, and decomposition, extracts low-frequency fault signatures embedded within excited higher-frequency resonances [14,15]. In addition to introducing adaptive demodulation band selection, Hilbert–Huang–based demodulation, and cyclostationary estimators have advanced the extraction of repetitive transient patterns under realistic industrial operation [16].

The analysis of bearing data and detection of possible faults is a time-dependent challenge, consisting of a time series classification problem. In recent years, ML and Deep Learning (DL) methods have been increasingly applied to tackle this challenge [5,17]. These techniques have proven to be effective in extracting meaningful patterns from complex vibration data, offering a powerful tool for fault diagnosis.

A wide range of methodologies has been developed over the past decade to address bearing fault detection, starting from more traditional machine learning techniques based on decision trees [18,19], support vector machines (SVMs) [20,21] and K-Nearest Neighbors (KNN) [22,23], to more recent methods, in the DL area, like CNNs [24,25], LSTMs [26,27], and Recurrent Neural Networks (RNNs) [28] to improve diagnostic accuracy and robustness. The main characteristics and results of these studies are summarized in Table 1.

Table 1. State-of-the-art in bearing fault classification.

Reference	Methodology	Key Features Used	Dataset	Accuracy
Traditional Machine Learning Methods				
Wan et al. (2021) [18]	Random Forest	Time–frequency-domain features	CWRU Dataset [29]	98.12%
Hang et al. (2019) [19]	Random Forest	Time-domain features	IEEE PHM 2012 Data Challenge Dataset [30]	93.4%
Janjarasjitt (2025) [20]	SVM	Time–frequency-domain features	CWRU Dataset [29]	100%
Jabbar et al. (2025) [21]	SVM, KNN	Time–frequency-domain features	MOIRA-UNIMORE Dataset [31], Politecnico di Torino Dataset [32]	100%
Wang et al. (2020) [22]	KNN	Time-domain features	CWRU Dataset [29]	96.1%
Lu et al. (2021) [23]	KNN	Time–frequency-domain features	Proprietary Dataset (Constant Speed)	96.67%
Deep Learning Methods				
Zhao et al. (2020) [24]	CNN	Time-domain features	CWRU Dataset [29], IMS Dataset [33]	99.2%
Zheng et al. (2024) [25]	EMDOS-DCCNN	Time-frequency-domain features	Proprietary Dataset (Variable Speed)	98.6%

Table 1. *Cont.*

Reference	Methodology	Key Features Used	Dataset	Accuracy
Guo et al. (2023) [26]	ACNN-BiLSTM	Time-domain features	CWRU Dataset [29]	99.79%
Öcalan et al. (2025) [27]	LSTM	Time-domain features	Ottawa Dataset [34]	100%
Liu et al. (2018) [28]	RNN-based Autoencoder	Time-domain features	CWRU Dataset [29]	>99%

In this study, we also propose using MC Dropout [35] to estimate uncertainty in deep learning model predictions, building on techniques commonly used in explainable AI [36], to better understand the reliability of the model's predictions. MC Dropout is a form of approximate Bayesian inference that enables uncertainty estimation during the inference phase by applying dropout at test time, which is typically only used during training.

In standard neural networks, dropout randomly disables a fraction of neurons during training to prevent overfitting. However, by applying dropout during testing, the model effectively generates a distribution of predictions, allowing us to quantify uncertainty. Specifically, by performing multiple stochastic forward passes with dropout, we can approximate the predictive distribution and estimate the uncertainty in the model's predictions.

The main idea behind MC Dropout, as shown in Figure 1a is that by performing multiple forward passes with different dropout masks, we can treat these different predictions as samples from the posterior distribution of the model. This allows the model to not only output predictions, but also provide a measure of confidence.

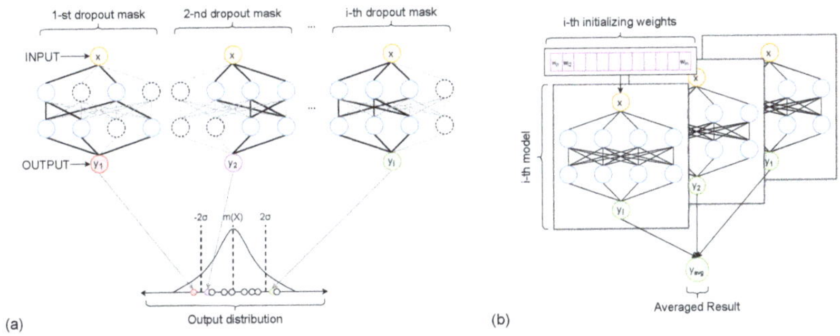

Figure 1. Uncertainty quantification methods. (**a**) **Monte Carlo Dropout**: x is the given input, and y_i are the outputs generated by applying n different dropout masks that form the final distribution. (**b**) **Deep Ensemble**: multiple models are independently trained with different initializations and combined at inference. The diversity among models enables improved accuracy and uncertainty estimation.

Ensemble learning [37] represents another widely used approach for estimating predictive uncertainty in deep learning models, commonly referred to as *Deep Ensembles*. Unlike MC Dropout, which relies on stochastic forward passes of a single neural network, Deep Ensembles employ multiple independently trained models, each initialized with different random weights and often trained using shuffled subsets of the training data [38]. The diversity among ensemble members arises from variations in initialization and training dynamics, which lead the models to converge to different local minima in the loss landscape. A graphical representation of this process is shown in Figure 1b.

Each model within the ensemble is typically trained using standard loss functions, such as Mean Squared Error (MSE) for regression tasks or Categorical Cross-Entropy for classification, without explicitly modeling uncertainty during training. In this standard setup, uncertainty estimation emerges naturally during inference, where the predictions of all ensemble members are aggregated, commonly by averaging their output probabilities, to obtain the final prediction. The variance among the individual model outputs reflects the *epistemic uncertainty* of the model: if all models yield similar predictions, the variance is low, indicating high confidence; conversely, large disagreement among models corresponds to higher uncertainty.

In more advanced formulations, probabilistic loss functions (e.g., those based on likelihood maximization or Bayesian principles) can be employed during training to capture not only epistemic but also *aleatoric uncertainty*—that is, uncertainty inherent in the data itself. However, such probabilistic training remains an extension beyond the standard Deep Ensemble framework.

3. Methodology

3.1. Experimental Setup

To create the dataset that we used in this paper, we used the ISE OneX Test Bench, shown in Figure 2a, which is a platform designed for simulating and evaluating the performance of rotating machines under different conditions, such as imbalances or defective components. The test bench includes a Siemens brushless AC motor powered by a 230 V, 50/60 Hz supply with a maximum current of 3 A. The motor's nominal power is 0.2 kW, with a standard speed of 3000 rpm and a maximum speed of 5000 rpm. Its nominal torque is 0.64 Nm (maximum 1.91 Nm), with three replaceable rolling bearings, identified (by the arrows) as 1,2,3 in Figure 2b, and two discs integral with the rotation axis. The test bench operates within a temperature range of +5 °C to +40 °C and can handle humidity levels up to 50% at +40 °C. For monitoring purposes, it is equipped with various sensors to track and analyze the machine's behaviors and performance degradation under various malfunctioning conditions.

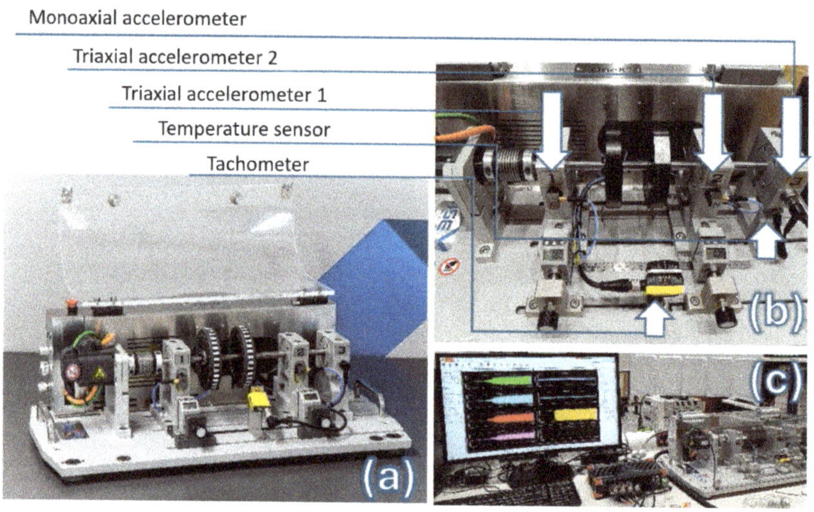

Figure 2. ISE OneX test-bench platform for evaluating the performance of rotating machines under different conditions (**a**). Integrated sensors with OneX test bench for monitoring (**b**). Experimental setup to measure the vibration using healthy and faulty bearings using 8-channel SIRIUSi data acquisition system (**c**).

As shown in Figure 2b, the test bench is equipped with different sensors. For vibration analysis, the bench includes two triaxial accelerometers manufactured by PCB PIEZOELECTRONICS placed on the bearings housing (1 and 2). These accelerometers have a sensitivity of 100 mV/g and are capable of detecting vibrations in all three axes. Additionally, an axial sensor from PCB PIEZOELECTRONICS with a sensitivity of 100 mV/g is used on bearing 3 for further vibration analysis. A retroreflective mode tachometer from Banner Engineering is used to monitor the speed of the rotating motor axis, providing precise speed control with a resolution of 1 rpm. A temperature sensor is placed on bearing 3 to measure temperature variations, with an accuracy of ±0.1 °C.

The test bench includes several test functionalities to simulate different mechanical issues and assess the system's response under various fault conditions. Imbalance simulation is achieved by adding or removing weights from the test bench's rotating disks, allowing for a test of the system's reaction to static and dynamic imbalances. Bearing wear is simulated by replacing healthy bearings with faulty bearings, enabling the analysis of system behavior when components are worn. The test bench also allows for the simulation of loose machine structures by adjusting its feet, which helps study the impact of loose components on performance. Additionally, both horizontal and vertical misalignments can be simulated by adjusting the positioning of the motor and bearings, enabling the evaluation of how misalignment affects machine operation. Figure 2c shows our experimental setup used to measure the vibration using healthy and faulty bearings. An 8-channel SIRIUSi data acquisition system, manufactured by DEWEsoft® (Trbovlje, Slovenia), is utilized to acquire signals. With the introduction of native integrated electronic piezoelectric (IEPE) mode, the module supports DC or AC mode input coupling with 0.16 Hz high-pass filter for low ranges (0.3 Hz for 100 V range). It features a dual 24-bit Delta-sigma ADC with a sampling rate of 200 kS/s for each channel. A summary of the sensors mounted on the testbench is shown in Table 2.

Table 2. Specifications of sensors and DAQ used in the experimental test bench. "–" are used when data is not available.

Sensor Type	Model/Manufacturer	Measured Quantity	Sensitivity/Range	Accuracy/Sampling Rate
Vibration Sensors				
Triaxial Accelerometer	PCB PIEZOELECTRONICS, Depew (NY), United States	Acceleration (X, Y, Z)	100 mV/g, ±50 g	20 kS/s
Axial Accelerometer	PCB PIEZOELECTRONICS, Depew (NY), United States	Acceleration (axial)	100 mV/g, ±50 g	20 kS/s
Rotational Speed Sensor				
Retroreflective Tachometer	Banner Engineering, Plymouth (MN)	Rotational speed (RPM)	– , 10 mm to 3 m polarized retroreflective sensing range	Resolution 1 RPM
DAQ System				
SIRIUSi Data Acquisition	DEWEsoft®, Trbovlje, Slovenia	Vibration, speed, temperature	–, –	24-bit Delta-sigma ADC/200 kS/s per channel

3.2. Dataset

The vibration signals analyzed in this work were acquired using two triaxial accelerometers, with a sampling frequency of 20 kHz, under different bearing conditions. The experimental campaign includes both healthy bearings and bearings with two distinct types of faults. The first fault type is the *Ball Pass Frequency Outer (BPFO)* race, which corresponds to localized damage on the outer race of the bearing. Such defects typically lead to periodic impacts every time a rolling element passes over the fault, producing strong and characteristic frequency components. The second fault type is the *Ball Spin Frequency (BSF)*, which arises from a defect on a rolling element (ball). In this case, the defect produces modulated vibration patterns, which are generally weaker and less periodic than BPFO, but still diagnostically relevant.

To train our models, two subsets of data were considered.

- **Constant speed:** this subset comprises 48 processes, of which 12 correspond to healthy bearings, 12 to BSF faults, and 24 to BPFO faults. During each process, the rotating speed was kept constant throughout the entire recording. The dataset includes 8 processes for each speed level in the range $[250, 5000]$ RPM, ensuring an even distribution across operating speeds.
- **Variable speed:** this subset extends the constant-speed dataset by including an additional 12 processes in which the rotational speed increases continuously from 0 to 5000 RPM. As shown in Figure 3, around 30 s were dedicated to each speed step. This configuration better reflects real-world scenarios, where rotating speed may vary during operation, thus providing the model with more realistic training data.

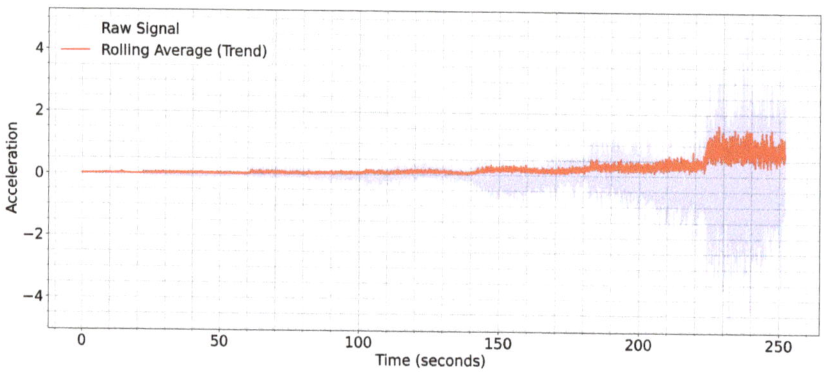

Figure 3. Visualization of speed steps in variable speed processes.

Each subset of the dataset was divided into training, validation, and test sets using an initial 60%-20%-20% split. Specifically, the constant speed phase included 28 processes for training, 10 for validation, and 10 for testing, while the variable speed phase comprised 36, 12, and 12 processes, respectively. This splitting procedure ensured that each subset was independent, with no process appearing in more than one subset. The original class distribution, in which the BPFO class contained twice as many samples as either the BSF or healthy classes, was preserved. To mitigate the effects of this class imbalance, a second, balanced version of the dataset was generated by downsampling the majority class to match the number of samples in the minority classes. This approach facilitates model training under conditions that minimize bias toward any particular fault type.

3.3. Feature Selection

The selection of features used as input for our models was performed gradually, with the goal of improving performance, and was guided by the specific challenges of fault detection and classification, as detailed in Section 4. The process leading to the final feature set can be divided into three main stages: (i) acceleration-only features, (ii) acceleration with envelope information, and (iii) statistical and spectral features in the time and frequency domains.

As described in Section 3.1, the testbench used for the creation of the dataset is equipped with two triaxial accelerometers. Consequently, the first set of features consisted of raw vibrational data along the x, y, and z axes for both bearings, capturing acceleration (m/s ^{2}) at each sampling instant, as can be seen in Figure 4. The raw data were segmented into windows of length 1000, resulting in an input tensor of shape $[1000, 6]$, where the six channels correspond to the three axes from each sensor.

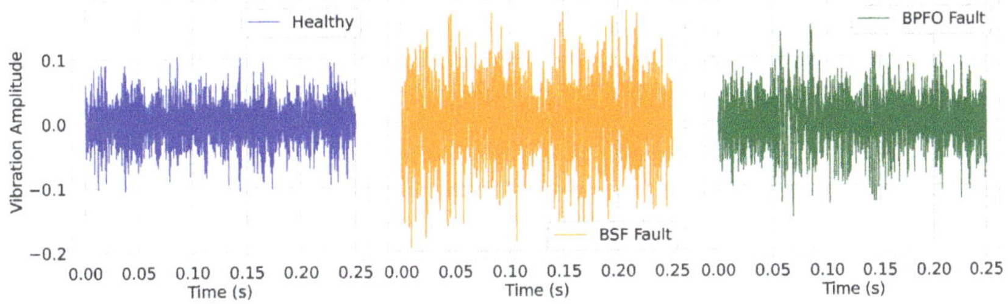

Figure 4. Acceleration data for a sample of each class in the dataset.

While raw acceleration captures instantaneous vibrations, it is often insufficient to fully characterize fault-related phenomena, as different fault types may not be clearly distinguishable in the raw time-domain signal. To address this, the second feature set includes the signal envelope for each axis and each bearing, thereby providing information on amplitude modulations, as shown in Figure 5. The envelope of a signal $x(t)$ is typically obtained by means of the analytic signal, defined as

$$x_a(t) = x(t) + j\mathcal{H}\{x(t)\}, \tag{1}$$

where $\mathcal{H}\{\cdot\}$ denotes the Hilbert transform. The envelope $e(t)$ is then

$$e(t) = |x_a(t)| = \sqrt{x(t)^2 + \mathcal{H}\{x(t)\}^2}. \tag{2}$$

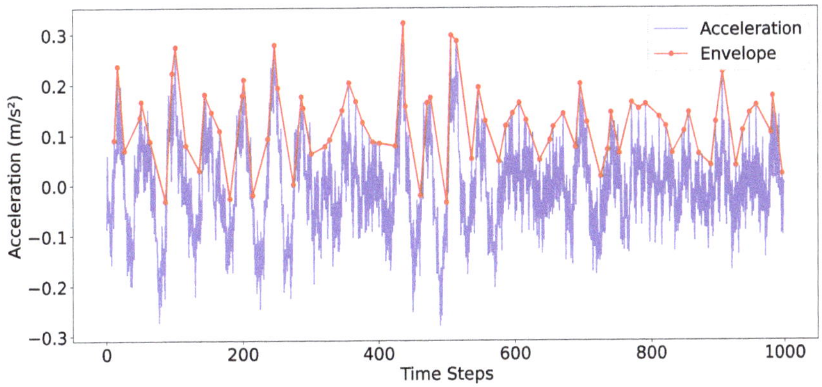

Figure 5. Example of acceleration and envelope on one axis.

This smooth curve outlines the amplitude variations over time, effectively suppressing the high-frequency oscillations. Studying the envelope is crucial for condition monitoring because many mechanical anomalies (e.g., bearing defects, misalignments) manifest primarily as changes in amplitude rather than frequency [39,40]. The inclusion of these features doubled the dimensionality of the input, yielding a final tensor of shape $[1000, 12]$.

To further improve model accuracy, we carried out an additional round of feature engineering aimed at identifying features with higher discriminative power. In this step, we explored non-linear transformations in both the time and frequency domains, while time-domain features capture overall signal behavior through statistical descriptors, the frequency domain is particularly important in ball bearing fault prediction because defects often generate characteristic vibration patterns at specific frequencies (e.g., fault frequencies linked to bearing geometry). These patterns are not always evident in the raw time signal but become more distinct once transformed into the frequency spectrum.

As a result, the final feature set combines time-domain statistical descriptors with frequency-domain characteristics, allowing us to better highlight differences between healthy and faulty conditions, as illustrated in Figure 6.

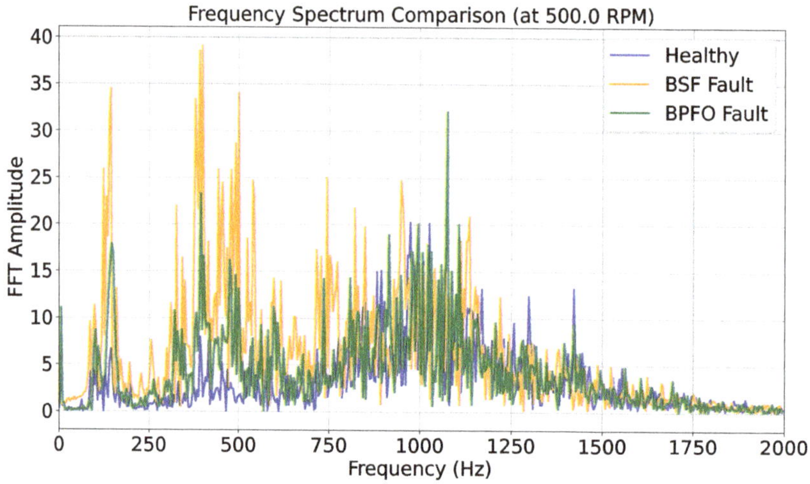

Figure 6. Frequency between 0 and 2000 Hz for each class in the dataset.

For each axis, the following features were computed:

- **Root Mean Square (RMS):** quantifies the signal's power, defined for a window $x[n]$, $n = 1, \ldots, N$ as

$$\mathrm{RMS} = \sqrt{\frac{1}{N}\sum_{n=1}^{N} x[n]^2}. \tag{3}$$

  An increase in RMS typically indicates higher energy in the vibration, often associated with faults.

- **Crest Factor (CF):** ratio of the maximum amplitude to the RMS, highlighting impulsive events:

$$\mathrm{CF} = \frac{\max|x[n]|}{\mathrm{RMS}}. \tag{4}$$

- **Kurtosis (K):** a normalized fourth central moment, capturing the "peakedness" of the distribution:

$$K = \frac{\frac{1}{N}\sum_{n=1}^{N}(x[n]-\mu)^4}{\sigma^4}, \tag{5}$$

  where μ and σ are the mean and standard deviation, respectively. High kurtosis is a strong indicator of early-stage bearing faults.

- **Skewness (*S*):** a measure of distribution asymmetry, given by

$$S = \frac{\frac{1}{N}\sum_{n=1}^{N}(x[n]-\mu)^3}{\sigma^3}. \quad (6)$$

Deviations from zero may signal asymmetric defect-induced vibrations.
- **Spectral Peaks:** the ten dominant frequency components were extracted from the magnitude spectrum of the Fast Fourier Transform (FFT). For each segment of length $N = 5000$, the FFT is computed as

$$X[k] = \sum_{n=0}^{N-1} x[n]e^{-j2\pi kn/N}, \quad k = 0,\ldots,N-1, \quad (7)$$

and the largest ten magnitudes $|X[k]|$ are retained as features, representing recurring periodicities linked to mechanical faults.

Combining these features results in 14 features per axis, leading to a total of 84 features per window across the six channels. A summary of all features used during the three steps is shown in Table 3.

Table 3. Summary of extracted features for bearing fault diagnosis.

Feature Name	Definition/Formula	Rationale
1st step		
Acceleration	Acceleration for both bearings on the 3 axis (m/s^2)	Captures raw vibrational energy and transient impulses caused by faults.
2nd step		
Hilbert Envelope	$e(t) = \lvert x_a(t)\rvert = \sqrt{x(t)^2 + \mathcal{H}\{x(t)\}^2}$	Highlights modulation produced by bearing fault impact forces.
3rd step		
RMS	$\sqrt{\frac{1}{N}\sum_{n=1}^{N} x[n]^2}$	Sensitive to impulsiveness and non-Gaussian behavior.
Crest Factor	$\frac{\max\lvert x[n]\rvert}{\text{RMS}}$	Identifies high-amplitude shocks relative to signal energy.
Kurtosis	$\frac{\frac{1}{N}\sum_{n=1}^{N}(x[n]-\mu)^4}{\sigma^4}$	Indicates early-stage bearing faults.
Skewness	$\frac{1}{N}\sum\left(\frac{x(n)-\mu}{\sigma}\right)^3$	Indicates asymmetry in vibration signature from progressing faults.
Spectral Peaks	Max amplitude in characteristic defect frequency bands	Reveals BPFO, BPFI, BSF, and FTF-related harmonics.

3.4. Models

The modeling approach was designed to evolve in parallel with the feature selection process. The models were built gradually, starting with simpler constant-speed measurements, where operating conditions are more stable. Once satisfactory performance and reliability were achieved, variable-speed measurements were introduced, allowing the models to adapt to non-stationary dynamics and enabling the refinement of both the architecture and its hyperparameters.

First, a CNN trained exclusively on acceleration-based features was developed to establish a baseline architecture. CNNs are feed-forward neural networks that employ convolutional filters to learn local patterns within structured data. By sharing weights across spatial or temporal dimensions, they efficiently detect translation-invariant features while maintaining a manageable number of parameters. This makes them particularly effective for vibration signals, where localized oscillatory patterns often correspond to specific fault conditions. Nonetheless, their limited capacity to model long-term temporal dependencies reduces their suitability for scenarios involving non-constant system behaviour, in which local feature extraction alone is insufficient.

The CNN was subsequently retrained with the inclusion of the envelope features, and later with a balanced version of the dataset obtained through undersampling to ensure all fault classes were equally represented. At each stage, the model was fine-tuned to maintain stable training behavior and achieve consistent benchmark performance. This progressive refinement led to a CNN architecture composed of three 1D convolutional blocks, each followed by max pooling and dropout layers to reduce dimensionality and mitigate overfitting. The convolutional filters, with a kernel size of 5, were designed to capture local variations in the vibration signal. The extracted feature maps were then flattened and passed through a fully connected layer before classification via a softmax output layer. The final structure of the CNN model is summarized as follows:

- Conv1D (128 filters, kernel size 5) + MaxPooling1D (pool size 3) + Dropout (0.5)
- Conv1D (64 filters, kernel size 5) + MaxPooling1D (pool size 3) + Dropout (0.5)
- Conv1D (32 filters, kernel size 5) + MaxPooling1D (pool size 3) + Dropout (0.5)
- Flatten + Dense(32, ReLU) + Dropout (0.5)
- Output: Dense (*num_classes*, Softmax)

With the addition of variable speed runs, the modeling framework was extended to include a hybrid architecture combining convolutional and recurrent layers. The CNN–BiLSTM model first employs convolutional and pooling layers to extract and compress local features, followed by a bidirectional Long Short-Term Memory (BiLSTM) layer that captures temporal relationships in both forward and backward directions. In doing so, it addresses the baseline CNN's limitations by incorporating long-range temporal context that is necessary under non-constant operating conditions, combining it with spatially localized features extracted by convolutional layers. Dropout layers are used after both convolutional and recurrent stages to reduce overfitting. The structure of the CNN–BiLSTM model is summarized as follows:

- Conv1D (32 filters, kernel size 5) + MaxPooling1D (pool size 2)
- Conv1D (64 filters, kernel size 5) + MaxPooling1D (pool size 2) + Dropout (0.5)
- Bidirectional LSTM (64 units) + Dropout (0.5)
- Output: Dense (*num_classes*, Softmax)

These two architectures were selected to explore the benefits of combining convolutional and recurrent processing in a unified framework. Each model underwent a pre-training phase for hyperparameter tuning to identify the best-performing configuration.

Finally, to further enhance robustness and generalization, a Deep Ensemble learning strategy was applied to the best model identified in the previous step. Multiple instances of the model were trained with different random initializations, and their predictions were aggregated at inference time. Ensemble learning reduces variance, improves stability, and mitigates overfitting, while retaining the modeling advantages of the underlying architecture.

4. Results

This section presents the classification results obtained across all experimental settings. Alongside the proposed architectures, we also report results obtained using standard deep-learning baselines (LSTM, BiLSTM, ResNet, and Transformer) to provide a clear comparison and contextualize the gains achieved by the hybrid and convolutional approaches. To ensure a fair and unbiased comparison across all model families, we performed systematic hyperparameter tuning using Keras Tuner, applying the same search strategy and comparable search budgets (15 configurations, 30 epochs each) to both the baseline and proposed models. This procedure guarantees that each architecture is evaluated under well-optimized configurations, addressing concerns regarding model selection and hyper-parameter fairness. For each experimental setting, we highlight the most relevant trends and discuss how different model families respond to variations in signal representation, class distribution, and operating speed.

4.1. Classification

4.1.1. Stage 1: Constant-Speed Evaluation

Table 4 summarizes the performance obtained when training exclusively on acceleration signals under constant-speed conditions. The results reveal a clear hierarchy among the tested architectures. Pure recurrent models (LSTM and BiLSTM) exhibit the lowest performance, reflecting their limited ability to extract meaningful fault-related patterns solely from raw vibration data. The CNN model, by contrast, achieves substantially higher accuracy (87.2%), highlighting the importance of convolutional filtering for capturing local temporal dependencies. The Transformer model also demonstrates competitive performance (96.6%), indicating that attention-based mechanisms can effectively model long-range interactions even in the absence of additional signal transformations. The CNN-BiLSTM architecture achieves near-perfect classification (99.4%), benefiting from the complementary strengths of convolutional feature extraction and bidirectional temporal modeling. ResNet attains the highest accuracy (99.9%), confirming the effectiveness of deep residual representations for fault classification under stationary operating conditions.

Table 4. Accuracy comparison on *Constant Speed* using only acceleration. Best results are in bold.

Model	Accuracy [%]	Recall [%]	F1-Score [%]
CNN	87.2	87.1	0.872
LSTM	77.3	77.3	0.769
BiLSTM	85.1	85.1	0.854
CNN-BiLSTM	99.4	99.4	0.994
Transformer	96.6	96.6	0.966
ResNet	**99.9**	**99.9**	**0.998**

The inclusion of envelope features (Table 5) leads to substantial improvements for most architectures. CNN performance increases to 97.0%, illustrating the relevance of demodulated vibration information for fault detection. Both CNN-BiLSTM and ResNet achieve nearly perfect scores, underscoring their ability to capitalize on the enriched representation. Interestingly, the Transformer exhibits a decrease in performance compared with the raw-acceleration setting, suggesting that its attention mechanisms may not benefit as directly from envelope information or may require additional feature tuning.

Table 5. Accuracy comparison on *Constant Speed* using acceleration and envelope. Best results are in bold.

Model	Accuracy [%]	Recall [%]	F1-Score [%]
CNN	97.0	97.1	0.970
LSTM	82.2	82.2	0.828
BiLSTM	83.0	83.0	0.831
CNN-BiLSTM	99.5	99.5	0.995
Transformer	90.8	90.8	0.906
ResNet	**99.9**	**99.9**	**0.999**

When the dataset is balanced to mitigate class imbalance (Table 6), its impact on the models is heterogeneous. CNN and ResNet maintain near-perfect accuracy, demonstrating strong generalization capability regardless of class distribution. In contrast, recurrent models (LSTM and BiLSTM) experience a significant degradation in performance, dropping to approximately 60% accuracy. This indicates a strong dependence on the original class priors and a limited ability to generalize when these priors are equalized, but also the need for more training data with respect to convolutional models. A plausible explanation for this drop is that recurrent models rely more strongly on temporal dependencies, which become less distinctive once class frequencies are equalized, rendering them more sensitive to changes in data distribution. CNN-BiLSTM also suffers a notable decrease, although it still outperforms the purely recurrent architectures.

Table 6. Accuracy comparison on *Constant Speed* after balancing the dataset. Best results are in bold.

Model	Accuracy [%]	Recall [%]	F1-Score [%]
CNN	**99.9**	99.8	0.998
LSTM	60.9	60.9	0.592
BiLSTM	61.2	61.2	0.609
CNN-BiLSTM	88.7	88.7	0.875
Transformer	76.0	76.0	0.766
ResNet	99.8	**99.8**	**0.998**

4.1.2. Stage 2: Variable-Speed Evaluation

The introduction of variable-speed data (0–5000 RPM) poses a more challenging classification scenario due to nonstationary effects. As shown in Table 7, all models exhibit reduced performance compared with the constant-speed case. Nonetheless, CNN-BiLSTM remains the best-performing architecture (93.7%), demonstrating resilience to temporal variability and speed-dependent spectral shifts. CNN also maintains strong results, whereas ResNet, despite its excellent performance under constant-speed conditions, shows a more pronounced drop, suggesting that its deep convolutional filters are less invariant to speed-induced distortions. Transformer and LSTM again yield the lowest accuracy values, reflecting limited adaptability to highly variable input dynamics.

Table 7. Accuracy comparison on *Variable Speed* using acceleration and envelope. Best results are in bold.

Model	Accuracy [%]	Recall [%]	F1-Score [%]
CNN	92.8	92.8	0.925
LSTM	78.4	78.4	0.776
BiLSTM	84.9	84.9	0.843
CNN-BiLSTM	**93.7**	**93.7**	**0.935**
Transformer	75.3	75.3	0.760
ResNet	88.7	88.7	0.884

4.1.3. Stage 3: Feature Engineering and Ensemble Learning

The adoption of advanced time- and frequency-domain descriptors (Table 8) significantly enhances performance across nearly all architectures. CNN attains 95.8% accuracy, confirming the benefit of engineered features instead of raw time-series representations. BiLSTM and especially CNN-BiLSTM show substantial gains, with the latter achieving 99.0%, indicating that hybrid architectures can effectively integrate both learned and hand-crafted representations. ResNet also improves markedly, reaching 98.1%, which suggests that engineered spectral indicators mitigate the sensitivity of convolutional filters to speed-induced variability.

The Deep Ensemble approach, consisting of training five CNN-BiLSTM models and averaging their outputs, achieves the highest overall performance, reaching 99.3% accuracy, recall, and F1-score. This result highlights the advantages of model averaging in reducing variance, increasing robustness, and producing more stable predictions under real-world, variable-speed operating conditions.

Overall, the results clearly demonstrate that convolution-based architectures, especially the CNN-BiLSTM approach, when combined with engineered features or ensemble learning offer superior generalization and robustness across both constant-speed and variable-speed operating conditions.

Table 8. Accuracy comparison on *Variable Speed* using advanced feature extraction. Best results are in bold.

Model	Accuracy [%]	Recall [%]	F1-Score [%]
CNN	95.8	95.7	0.955
LSTM	78.9	79.0	0.790
BiLSTM	91.1	91.1	0.910
CNN-BiLSTM	99.0	98.7	0.987
Transformer	83.0	82.9	0.830
ResNet	98.1	98.1	0.981
Deep Ensemble	**99.3**	**99.3**	**0.993**

4.2. Uncertainty Estimation

To enhance the interpretability and reliability of the model's predictions, we implemented two complementary approaches to quantify uncertainty: **MC Dropout** and **Deep Ensemble** learning. Both methods allow the user to go beyond single-point predictions by providing not only an expected output but also a measure of confidence in the prediction.

MC Dropout involves enabling dropout during inference, a mechanism typically used only during training. By performing multiple stochastic forward passes for the same input, the network effectively produces a distribution of outputs. In our experiments, we generated 100 predictions per input sample. Each forward pass randomly deactivates a fraction of neural connections, creating slightly different models and thus capturing model uncertainty. From these 100 outputs, we compute the mean prediction $\bar{y}$ and the standard deviation σ, representing the confidence of the network. Correct predictions generally show a low standard deviation (high confidence), while incorrect or ambiguous predictions exhibit higher dispersion. This allows end users to define thresholds on σ to trigger alarms or request human intervention, mitigating the risk of blind reliance on the model.

As illustrated in Figure 7, the uncertainty metric effectively differentiates between reliable and uncertain predictions. For a correct classification, the mean prediction is strongly concentrated on the correct class, and the standard deviation σ is nearly zero, signifying high confidence and robustness. Conversely, for a misclassified sample, the mean prediction $\bar{y}$ is distributed across multiple classes (BSF has the highest mean probability, but is not the true class), and the large vertical error bars clearly indicate a high standard deviation σ. This correlation between high σ and misclassification confirms the value of using uncertainty to enhance the trustworthiness of the AI system, particularly when dealing with complex, non-stationary data.

In addition to MC Dropout, we applied a Deep Ensemble strategy, combining predictions from multiple independently trained CNN-BiLSTM models. Each model produces its own output for a given input, and the ensemble prediction is obtained by averaging these outputs. Similar to MC Dropout, the standard deviation across the ensemble serves as an uncertainty measure, highlighting cases where the models disagree.

The ability of Deep Ensembles to provide reliable measures of predictive uncertainty is demonstrated by examining the probability distributions of predictions for correctly classified and misclassified samples. Figures 8 and 9 present boxplots of the ensemble prediction probabilities under these two scenarios, allowing for a direct comparison of confidence levels and dispersion across classes.

Figure 8 illustrates the prediction distributions when the ensemble correctly identified the true class. In these cases, the predicted probability for the correct class is highly concentrated, with the median probability lying effectively at 1.0 across all three true classes. This indicates an ensemble prediction of high confidence. Moreover, the boxplots for the true class are extremely narrow, revealing minimal variance in the predictions among ensemble members. Such a low level of dispersion demonstrates consistent agreement across the ensemble and reflects an extremely low epistemic uncertainty. In practice, this consistency signifies that the ensemble classification is not only correct but also highly robust and trustworthy.

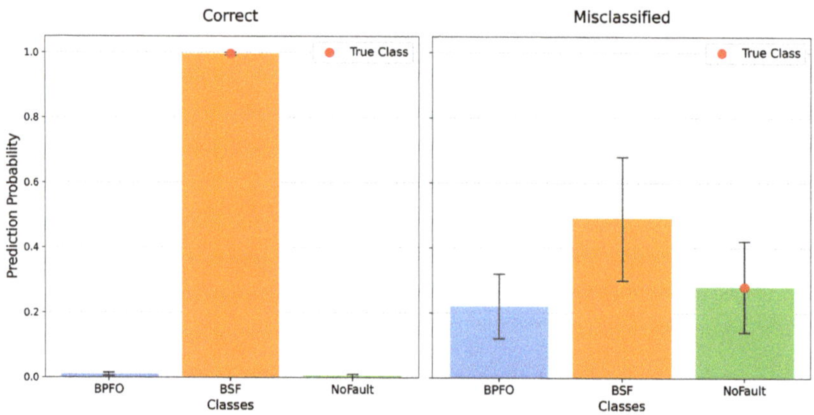

Figure 7. Predictions made using MC Dropout approach on two test samples.

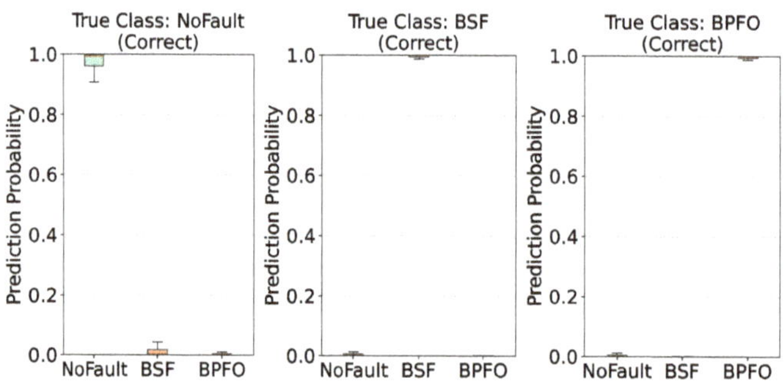

Figure 8. Uncertainty distribution in correctly classified samples for each class using Deep Ensemble.

In contrast, Figure 9 depicts prediction distributions for samples that the ensemble misclassified. Unlike the sharp concentration observed for correct classifications, the distributions here display a much wider spread, with visibly larger interquartile ranges and greater overall variability. This increased dispersion corresponds to high predictive uncertainty, arising from disagreement among the ensemble members. Thus, high variance effectively serves as a built-in alert mechanism, signalling instances where the model is less reliable and where human review may be required.

Another key feature of the misclassified cases is the splitting of probability mass across multiple classes. For example, in the case of samples truly belonging to the *BSF* class but incorrectly classified, the ensemble's probability mass was concentrated instead around the *NoFault* class, with a median predicted probability of approximately 0.63. Similarly, for true *BPFO* samples, the ensemble often assigned the highest probability to *NoFault*, with a median around 0.61. In both cases, the true class probability was not only lower but also subject to greater uncertainty (e.g., *BSF* samples had a median predicted probability for their true class of only 0.25). Such behaviour highlights how uncertainty manifests in practical error cases: the ensemble struggles to reach consensus, spreading its predictions across competing classes.

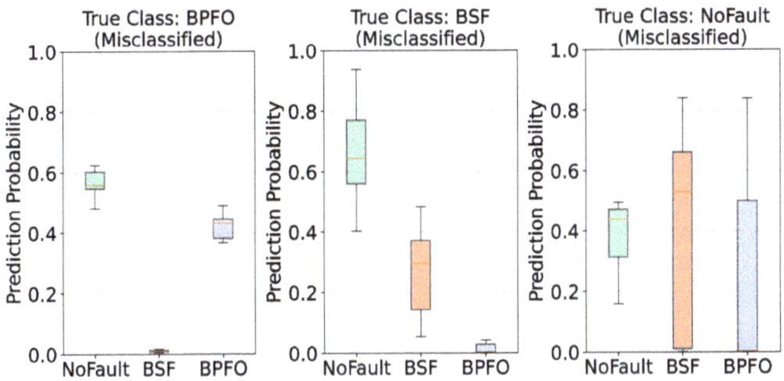

Figure 9. Uncertainty distribution in misclassified samples for each class using Deep Ensemble.

4.3. Explainable AI

To further increase the transparency of the model's decision-making process, moving the system toward a "grey-box" approach, we utilized Shapley Additive Explanations (SHAP). Based on cooperative game theory, SHAP is a model-agnostic method that assigns an importance value (the SHAP value) to each feature, quantifying its contribution (positive or negative) to the final prediction for a given sample, relative to the average model output.

The SHAP analysis provides global insights into which features were most influential in achieving the final ensemble accuracy of 99.3% (Table 8). The results are presented in two summary plots, grouping features by bearing and by feature type.

Figure 10 provides a high-level comparison of the feature importance aggregated by the sensor location: *ACC1 All Features* (Bearing 1) versus *ACC2 All Features* (Bearing 2). In a SHAP summary plot, features are ranked by the average magnitude of their impact, and the color represents the feature's actual value (red for high, blue for low).

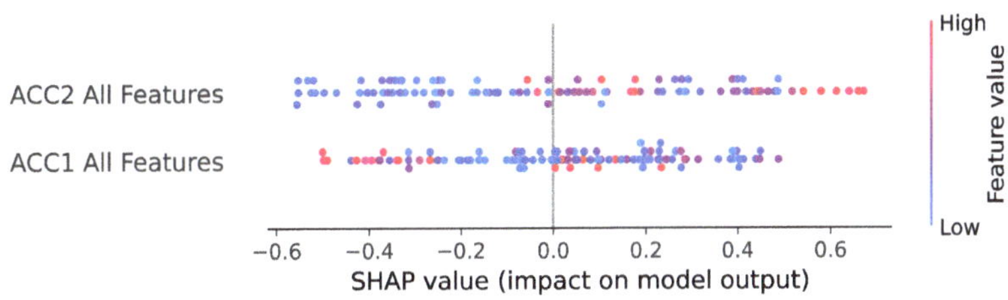

Figure 10. SHAP analysis at bearing level.

The plot indicates that, across the entire dataset, features from *ACC2 All Features* have a consistently greater overall impact on the model's prediction than those from *ACC1*. This suggests that the vibrational characteristics measured at the second bearing position were slightly more discriminative for fault diagnosis; indeed, it is where the faulty bearing was placed in our measurements.

A more granular breakdown is presented in Figure 11, which groups the features into their specific domain categories: *Time Features* (RMS, Crest Factor, kurtosis, skewness) and *Frequency Features* (spectral peaks).

This analysis strongly validates the feature engineering strategy detailed in Section 3.3 and the performance results in Section 4.1:

- **Dominance of Frequency Features:** the *ACC1 Frequency Features* and *ACC2 Frequency Features* groups exhibit the largest SHAP value magnitudes. This confirms that the spectral components, which directly correspond to the physical fault frequencies (BPFO and BSF), are the primary drivers of the model's high diagnostic accuracy. The model correctly prioritizes these domain-expert-derived features to distinguish between different fault types.
- **Role of Time-Domain Features:** while significant, the *ACC1 Time Features* and *ACC2 Time Features* groups show a comparatively lower overall impact. This indicates that traditional time-domain statistical descriptors, such as kurtosis, play a necessary but secondary role. They likely contribute to detecting the existence of an impulsive event (early-stage fault detection) and general severity (via RMS), but the frequency features are essential for accurate classification of the fault type.

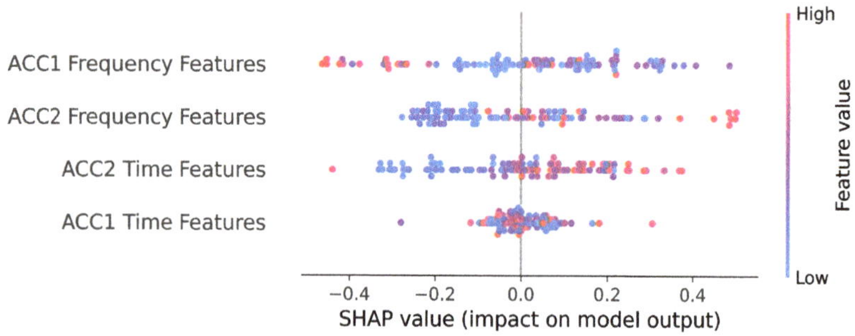

Figure 11. SHAP analysis at feature level.

By correlating the high SHAP values associated with the frequency-domain features with the spectral density analysis shown in Figures 12 and 13, restricting the comparison to processes operating at 500 RPM to ensure coherence, we observe that the model's decision boundary is primarily influenced by signal magnitude within the structural resonance bands. Figure 12 reveals a multimodal distribution shared by all three classes in the 500 Hz and 800–1000 Hz regions, reflecting the inherent structural resonances of the machinery. In contrast, the faulty classes exhibit an additional spectral component in the 300–400 Hz band that is absent in the healthy state.

Although the frequency locations of the dominant peaks substantially overlap across classes, Figure 13 indicates that signal amplitude serves as the key discriminative factor. The healthy class is characterized by low-magnitude vibrations across all resonance bands, whereas both fault conditions display a pronounced increase in spectral energy. In particular, the *BSF* class shows the highest concentration of spectral density in the 500 Hz band, suggesting that this fault induces a narrowband excitation that strongly couples with the machine's natural frequency. Conversely, the *BPFO* class exhibits a broader distribution of high-energy peaks within the 400–600 Hz range, consistent with the more impulsive, wide-band excitation typically associated with outer-race defects.

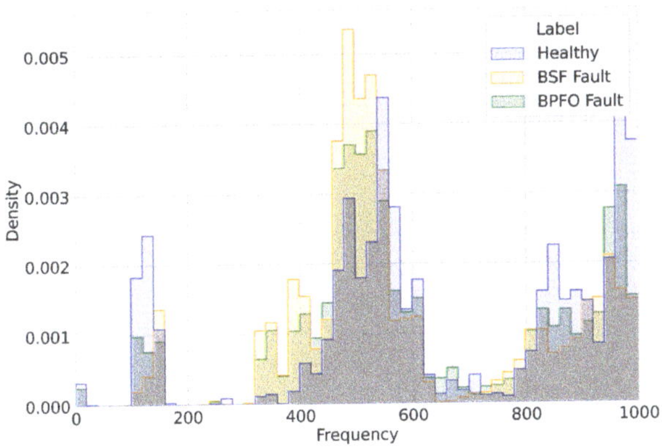

Figure 12. Analysis of dominant frequencies in processes operating at 500 RPM.

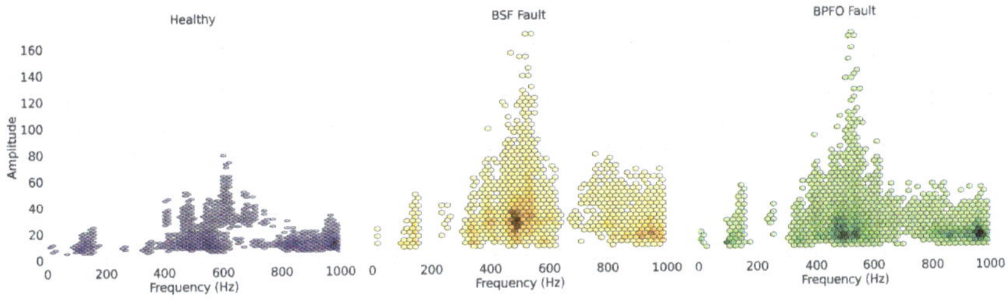

Figure 13. Analysis of signal amplitude in processes operating at 500 RPM.

4.4. Model Training and Inference Performance

Table 9 summarizes the computational characteristics of the evaluated models on the extended dataset using the complete feature set, reporting their training time, inference time, and number of trainable parameters. Overall, ResNet and the CNN variants, show the lowest computational cost, fast inference and relatively small parameter counts. Recurrent models exhibit higher training and inference times, particularly the bidirectional variants, which roughly double the computational load compared to their unidirectional counterparts. The Transformer model presents the highest training time among the deep architectures, reflecting its increased complexity despite a moderate parameter count.

Table 9. Summary of model computational metrics.

Model	Training Time	Inference Time	N. of Parameters
CNN	3 min 4 s	7 ms	56,259
CNN-biLSTM	1 min 58 s	8 ms	76,931
LSTM	3 min 0 s	12 ms	30,467
biLSTM	5 min 0 s	23 ms	77,187
Transformer	19 min 26 s	86 ms	64,483
ResNet	55 s	4 ms	51,555

5. Conclusions

In this study, we introduced a new rolling bearing vibration dataset that captures healthy and faulty conditions under both constant and variable speeds, including ball pass frequency outer race and ball spin frequency defects. Using this dataset, we developed a CNN-based fault classification framework enhanced with white-box feature selection, incorporating raw acceleration, envelope, and time- and frequency-domain statistical descriptors.

Experimental results demonstrate that raw acceleration signals alone provide sufficient information to distinguish healthy from faulty bearings (87.2% accuracy) under constant speed, but additional features are essential for accurately classifying fault types. Incorporating envelopes improves performance conditions (97.0%), while balancing the dataset further increases accuracy to 99.9%.

Under variable-speed conditions, classification complexity increases, highlighting the need for robust architectures; a CNN-BiLSTM hybrid with advanced features achieves 98.7%, and a Deep Ensemble strategy reaches 99.3% accuracy.

Moreover, MC Dropout and Deep Ensemble effectively quantifies prediction uncertainty, offering confidence measures that can guide maintenance decisions. SHAP analysis was used to confirm the dominance of frequency-domain features in fault discrimination, validating the grey-box approach. Overall, the proposed methodology provides a robust, interpretable, and high-performing framework for industrial predictive maintenance.

Future research will pursue several complementary directions aimed at enhancing the performance and generalization capabilities of the proposed model. A primary focus will be the development of an improved version of the dataset, incorporating more heterogeneous data, including additional defect types as well as the same defects exhibited at varying levels of severity. This expanded dataset is expected to enable the model to capture subtle variations in defect patterns and improve robustness across a wider spectrum of operational conditions. Subsequently, the model will be evaluated under real-world conditions, providing critical insights into the model's ability to generalize beyond the controlled laboratory environment used during training.

On the methodological front, future work will explore alternative deep learning architectures, including attention-based models beyond the transformer architectures already tested. Additionally, domain adaptation techniques, such as Domain-Adversarial Neural Networks (DANNs), will be investigated to facilitate knowledge transfer across datasets with differing characteristics, improving performance especially on smaller datasets, tackling the problem of costly data acquisition campaigns.

Author Contributions: Conceptualization, L.M., P.E., and L.C.; Methodology L.M., P.E., and A.M.; Supervision L.C. All authors have read and agreed to the published version of the manuscript.

Funding: This study was carried out within the MICS (Made in Italy—Circular and Sustainable) Extended Partnership and received funding from Next-Generation EU (Italian PNRR—M4 C2, Invest 1.3—D.D. 1551.11-10-2022, PE00000004). CUP MICS D43C22003120001.

Data Availability Statement: The raw data supporting the conclusions of this article will be made available by the authors on request.

Conflicts of Interest: The authors declare no conflicts of interest.

References

1. Cristaldi, L.; Esmaili, P.; Gruosso, G.; La Bella, A.; Mecella, M.; Scattolini, R.; Arman, A.; Susto, G.A.; Tanca, L. The mics project: A data science pipeline for industry 4.0 applications. In Proceedings of the 2023 IEEE International Conference on Metrology for eXtended Reality, Artificial Intelligence and Neural Engineering (MetroXRAINE), Milano, Italy, 25–27 October 2023; IEEE: New York, NY, USA, 2023; pp. 427–431.
2. Vashishtha, G.; Chauhan, S.; Sehri, M.; Zimroz, R.; Dumond, P.; Kumar, R.; Gupta, M.K. A roadmap to fault diagnosis of industrial machines via machine learning: A brief review. *Measurement* **2025**, *242*, 116216. [CrossRef]
3. Rehman, A.U.; Jiao, W.; Jiang, Y.; Wei, J.; Sohaib, M.; Sun, J.; Rehman, K.U.; Chi, Y. Deep learning in industrial machinery: A critical review of bearing fault classification methods. *Appl. Soft Comput.* **2025**, *171*, 112785. [CrossRef]
4. Afshar, M.; Heydarzadeh, M.; Akin, B. A comprehensive investigation of fault signatures and spectrum analysis of vibration signals in distributed bearing faults. *IEEE Trans. Ind. Appl.* **2024**, *61*, 515–526. [CrossRef]

5. Martiri, L.; Esmaili, P.; Cristaldi, L. A Novel Benchmark for Fault Detection in Rolling Bearings Using CNNs and Monte Carlo Dropout. In Proceedings of the 2025 IEEE International Workshop on Metrology for Industry 4.0 & IoT (MetroInd4.0 & IoT), Castelldefels, Spain, 1–3 July 2025; pp. 462–466.
6. Esmaili, P.; Cristaldi, L. Health indicator analysis in terms of condition monitoring on brownfield cnc milling machines using triaxial accelerometer. *IEEE Sens. Lett.* **2024**, *8*, 6008404. [CrossRef]
7. Randall, R.B.; Antoni, J. Rolling element bearing diagnostics—A tutorial. *Mech. Syst. Signal Process.* **2011**, *25*, 485–520. [CrossRef]
8. Althubaiti, A.; Elasha, F.; Teixeira, J.A. Fault diagnosis and health management of bearings in rotating equipment based on vibration analysis—A review. *J. Vibroeng.* **2022**, *24*, 46–74. [CrossRef]
9. Esmaili, P.; Cristaldi, L. Health indicator effectiveness in localized fault diagnosis: Rolling bearing elements. In Proceedings of the 2024 IEEE International Workshop on Metrology for Industry 4.0 & IoT (MetroInd 4.0 & IoT), Firenze, Italy, 29–31 May 2024; pp. 552–556.
10. Antoni, J. Cyclostationarity by examples. *IEEE Trans. Instrum. Meas.* **2016**, *65*, 1256–1264. [CrossRef]
11. He, W.; Shen, Q.; Sun, S. Envelope analysis-based diagnosis of bearing defects under variable speed conditions. *IEEE Trans. Ind. Electron.* **2016**, *63*, 5682–5692.
12. Song, S.; Wang, W. Early Fault Detection of Rolling Bearings Based on Time-Varying Filtering Empirical Mode Decomposition and Adaptive Multipoint Optimal Minimum Entropy Deconvolution Adjusted. *Entropy* **2023**, *25*, 1452. [CrossRef]
13. Sandoval Núñez, D.A.; Leturiondo, U.; Vidal, Y.; Pozo, F. Entropy Indicators: An Approach for Low-Speed Bearing Diagnosis. *Sensors* **2021**, *21*, 849. [CrossRef]
14. Yang, X.; Yang, J.; Jin, Y.; Liu, Z. A New Method for Bearing Fault Diagnosis across Machines Based on Envelope Spectrum and Conditional Metric Learning. *Sensors* **2024**, *24*, 2674. [CrossRef] [PubMed]
15. Ruiz-Sarrio, J.E.; Antonino-Daviu, A.; Martís, C. Localized Bearing Fault Analysis for Different Induction Machine Start-Up Modes via Vibration Time–Frequency Envelope Spectrum. *Sensors* **2024**, *24*, 6935. [CrossRef] [PubMed]
16. Tang, S.; Yuan, S.; Zhu, Y. Cyclostationary Analysis towards Fault Diagnosis of Rotating Machinery: Review, Advances, and Challenges. *Processes* **2020**, *8*, 1217. [CrossRef]
17. Martiri, L.; Esmaili, P.; Cristaldi, L. A Convolutional Neural Network for CNC Milling Machines Processes Classification. In Proceedings of the 2024 IEEE International Conference on Metrology for eXtended Reality, Artificial Intelligence and Neural Engineering (MetroXRAINE), St Albans, UK, 21–23 October 2024; IEEE: New York, NY, USA, 2024; pp. 634–639.
18. Wan, L.; Gong, K.; Zhang, G.; Yuan, X.; Li, C.; Deng, X. An efficient rolling bearing fault diagnosis method based on spark and improved random forest algorithm. *IEEE Access* **2021**, *9*, 37866–37882. [CrossRef]
19. Hang, Q.; Yang, J.; Xing, L. Diagnosis of rolling bearing based on classification for high dimensional unbalanced data. *IEEE Access* **2019**, *7*, 79159–79172. [CrossRef]
20. Janjarasjitt, S. Investigating the Effect of Vibration Signal Length on Bearing Fault Classification Using Wavelet Scattering Transform. *Sensors* **2025**, *25*, 699. [CrossRef]
21. Jabbar, A.; D'Elia, G.; Cocconcelli, M. Distribution Reshaping Transformation for Bearing Fault Diagnosis in Independent Cart Systems. *IEEE Access* **2025**, *13*, 200403–200430. [CrossRef]
22. Wang, H.; Yu, Z.; Guo, L. Real-time online fault diagnosis of rolling bearings based on KNN algorithm. In Proceedings of the 2019 4th International Seminar on Computer Technology, Mechanical and Electrical Engineering (ISCME 2019), Chengdu, China, 13–15 December 2019; Journal of Physics: Conference Series; IOP Publishing: Bristol, UK, 2020; Volume 1486, p. 032019.
23. Lu, Q.; Shen, X.; Wang, X.; Li, M.; Li, J.; Zhang, M. Fault diagnosis of rolling bearing based on improved VMD and KNN. *Math. Probl. Eng.* **2021**, *2021*, 2530315. [CrossRef]
24. Zhao, B.; Zhang, X.; Li, H.; Yang, Z. Intelligent fault diagnosis of rolling bearings based on normalized CNN considering data imbalance and variable working conditions. *Knowl.-Based Syst.* **2020**, *199*, 105971. [CrossRef]
25. Zheng, X.; Liu, X.; Zhu, C.; Wang, J.; Zhang, J. Fault diagnosis of variable speed bearing based on EMDOS-DCCNN model. *J. Vib. Eng. Technol.* **2024**, *12*, 7193–7207. [CrossRef]
26. Guo, Y.; Mao, J.; Zhao, M. Rolling bearing fault diagnosis method based on attention CNN and BiLSTM network. *Neural Process. Lett.* **2023**, *55*, 3377–3410. [CrossRef]
27. Öcalan, G.; Türkoğlu, İ. Diagnosis of Bearing Faults Under Variable Speed Conditions Using Deep Learning. *Veri Bilim.* **2025**, *8*, 1–10.
28. Liu, H.; Zhou, J.; Zheng, Y.; Jiang, W.; Zhang, Y. Fault diagnosis of rolling bearings with recurrent neural network-based autoencoders. *Isa Trans.* **2018**, *77*, 167–178. [CrossRef] [PubMed]
29. Bearing Data Center. Bearing Data Center Website. 2019. Available online: https://engineering.case.edu/bearingdatacenter (accessed on 13 November 2025).
30. Nectoux, P.; Gouriveau, R.; Medjaher, K.; Ramasso, E.; Chebel-Morello, B.; Zerhouni, N.; Varnier, C. PRONOSTIA: An experimental platform for bearings accelerated degradation tests. In Proceedings of the IEEE International Conference on Prognostics and Health Management, PHM'12, Denver, CO, USA, 18–21 June 2012; pp. 1–8.

31. Jabbar, A.; Cocconcelli, M.; D'Elia, G.; Borghi, D.; Capelli, L.; Cavalaglio Camargo Molano, J.; Strozzi, M.; Rubini, R. MOIRA-UNIMORE Bearing Data Set for Independent Cart Systems. *Appl. Sci.* **2025**, *15*, 3691. [CrossRef]
32. Daga, A.P.; Fasana, A.; Marchesiello, S.; Garibaldi, L. The Politecnico di Torino rolling bearing test rig: Description and analysis of open access data. *Mech. Syst. Signal Process.* **2019**, *120*, 252–273. [CrossRef]
33. Lee, J.; Qiu, H.; Yu, G.; Lin, J.; Services, R.T. Bearing Data Set. IMS, University of Cincinnati. NASA Prognostics Data Repository, NASA Ames Research Center, Moffett Field, CA. 2007. Available online: https://phm-datasets.s3.amazonaws.com/NASA/4.+Bearings.zip (accessed on 13 November 2025).
34. Huang, H.; Baddour, N. Bearing vibration data collected under time-varying rotational speed conditions. *Data Brief* **2018**, *21*, 1745–1749. [CrossRef]
35. Gal, Y.; Ghahramani, Z. Dropout as a bayesian approximation: Representing model uncertainty in deep learning. In Proceedings of the International Conference on Machine Learning. PMLR, New York, NY, USA, 19–24 June 2016; pp. 1050–1059.
36. Abdar, M.; Pourpanah, F.; Hussain, S.; Rezazadegan, D.; Liu, L.; Ghavamzadeh, M.; Fieguth, P.; Cao, X.; Khosravi, A.; Acharya, U.R.; et al. A review of uncertainty quantification in deep learning: Techniques, applications and challenges. *Inf. Fusion* **2021**, *76*, 243–297. [CrossRef]
37. Ganaie, M.A.; Hu, M.; Malik, A.K.; Tanveer, M.; Suganthan, P.N. Ensemble deep learning: A review. *Eng. Appl. Artif. Intell.* **2022**, *115*, 105151. [CrossRef]
38. Mohammed, A.; Kora, R. A comprehensive review on ensemble deep learning: Opportunities and challenges. *J. King Saud Univ. Comput. Inf. Sci.* **2023**, *35*, 757–774. [CrossRef]
39. Areias, I.A.d.S.; Borges da Silva, L.E.; Bonaldi, E.L.; de Lacerda de Oliveira, L.E.; Lambert-Torres, G.; Bernardes, V.A. Evaluation of current signature in bearing defects by envelope analysis of the vibration in induction motors. *Energies* **2019**, *12*, 4029. [CrossRef]
40. Randall, R.B.; Antoni, J.; Chobsaard, S. The relationship between spectral correlation and envelope analysis in the diagnostics of bearing faults and other cyclostationary machine signals. *Mech. Syst. Signal Process.* **2001**, *15*, 945–962. [CrossRef]

instruments

MDPI

Article

Shoulder Muscle Strength Assessment: A Comparative Study of Hand-Held Dynamometers and Load Cell Measurements

Carla Antonacci [1,2], Arianna Carnevale [2], Letizia Mancini [1,2], Alessandro de Sire [3,4], Pieter D'Hooghe [5], Michele Mercurio [4,6], Rocco Papalia [2,7], Emiliano Schena [1,2] and Umile Giuseppe Longo [2,7,*]

1 Research Unit of Measurements and Biomedical Instrumentation, Università Campus Bio-Medico di Roma, Via Alvaro del Portillo, 21, 00128 Roma, Italy; carla.antonacci@unicampus.it (C.A.); l.mancini@policlinicocampus.it (L.M.); e.schena@unicampus.it (E.S.)
2 Fondazione Policlinico Universitario Campus Bio-Medico, Via Alvaro del Portillo, 200, 00128 Roma, Italy; arianna.carnevale@policlinicocampus.it (A.C.); r.papalia@policlinicocampus.it (R.P.)
3 Department of Medical and Surgical Sciences, Research Center on Musculoskeletal Health, University of Catanzaro "Magna Graecia", 88100 Catanzaro, Italy; alessandro.desire@unicz.it
4 Research Center on Musculoskeletal Health, MusculoSkeletalHealt@UMG, Magna Græcia University, 88100 Catanzaro, Italy; michele.mercurio@unicz.it
5 Department of Orthopaedic Surgery and Sports Medicine, Aspetar Hospital, Doha 29222, Qatar; pieter.dhooghe@aspetar.com
6 Department of Orthopedic and Trauma Surgery, Magna Græcia University, Renato Dulbecco University Hospital, 88100 Catanzaro, Italy
7 Research Unit of Orthopaedic and Trauma Surgery, Department of Medicine and Surgery, Università Campus Bio-Medico di Roma, Via Alvaro del Portillo, 21, 00128 Roma, Italy
* Correspondence: g.longo@unicampus.it or g.longo@policlinicocampus.it

Abstract

Accurate measurement of shoulder muscle strength is important for diagnosis, treatment planning, and monitoring recovery. Hand-held dynamometers (HHDs) are widely used in clinical practice but are affected by operator strength, patient positioning, and device stabilization, particularly under high-load conditions. No previous study has directly compared HHD measurements with a reference load cell in a rigid serial configuration or evaluated the effect of different load cell signal processing strategies on the final strength value. The aim of this study was to compare HHD measurements with those obtained from a reference load cell in a rigid serial configuration and to assess how different signal processing strategies applied to load cell data influence the final outcomes. A custom 3D-printed support was developed to align a commercial HHD and a load cell in series, ensuring identical loading conditions. Measurements were performed under two conditions: (i) application of known weights (9.81–98.10 N) and (ii) standardized strength tasks in five healthy volunteers. Agreement between instruments was evaluated using Bland–Altman analysis and Root Mean Square Error (RMSE). In static validation (i.e., experiments applying know weights), the load cell demonstrated stable performance, with standard deviations below 1% of the applied load. HHD variability increased with load, with RMSE rising from 0.55 N at 9.81 N to 5.06 N at 98.10 N. In human testing, the HHD consistently underestimated muscle strength compared with the load cell, with mean differences ranging from −15 N to −19 N, over exerted force ranges of approximately 20–90 N. Overall, the load cell provided stable reference measurements, while the choice of signal processing strategy influenced the results: plateau-phase analysis tended to reduce systematic bias but did not consistently narrow the limits of agreement.

Keywords: shoulder; muscle strength assessment; hand-held dynamometer; load cell; clinical testing

1. Introduction

The shoulder is a complex and kinematically redundant joint, characterized by high mobility resulting from the coordinated interaction of three bones (clavicle, scapula, and humerus) and four joints (sternoclavicular, acromioclavicular, glenohumeral, and scapulothoracic) [1,2]. Under physiological conditions, shoulder motion is primarily governed by muscular action [1–4].

The assessment of shoulder muscle strength is essential for evaluating musculoskeletal function, supporting diagnosis, guiding treatment planning, and monitoring clinical recovery [2,5–7]. In clinical practice, several scoring systems are used to assess shoulder function, especially in pre- and post-treatment phases, to detect short- and long-term changes in symptoms and disabilities [6,8,9]. Among these, the Constant Murley Score is one of the most widely used assessments [10–13]. It integrates measures of pain, range of motion, daily living activities, and muscle strength, thus providing a comprehensive evaluation of shoulder function [10,11,14]. Muscle strength is a key element of the Constant Murley Score, and it has a significant impact on the final evaluation, frequently influencing future treatment choices [14,15].

Several methods and instruments have been developed to obtain objective measurements of shoulder muscle strength [16,17]. Hand-held dynamometers (HHDs) are commonly used due to their portability, ease of use, and low cost [7,16,18–21]. However, the measurements collected with this instrument may be affected by factors such as tester strength, subject positioning, and device stabilization [16–18,20–25]. These limitations become more evident during high-force tasks, where reduced stabilization can compromise measurement consistency [17,20,21,26–29]. Studies indicate that HHDs perform best under lower-force conditions, typically below 120 N, where stabilization challenges are minimized, while higher-force assessments, often exceeding 200 N, generally show reduced reliability and may require alternative stabilization methods or equipment to improve accuracy [17,30–33]. Given these limitations, it was crucial to evaluate the performance of HHD measurements across different testing conditions to ensure accuracy and consistency, especially for clinical assessments in patients with shoulder musculoskeletal disorders during relevant movements.

Recent studies have investigated advanced devices for assessing muscle strength, such as load cells, due to their high accuracy and data reliability [34–37]. However, research on their application for shoulder muscle strength evaluation remains limited. Existing evidence suggests that load cells may be less influenced by factors like subject or examiner anthropometric characteristics compared to HHDs [34]. Despite these advantages, their application in clinical settings remains limited, primarily due to setup complexity and the requirement for dedicated acquisition and processing software [34–37]. One limitation of load cells in clinical strength assessment is that they do not provide an immediate scalar output, such as peak or average force, that can be directly entered into scoring systems like the Constant-Murley Score. Instead, the output is a continuous force-time signal spanning the entire test duration. To obtain a usable value, the signal must be post-processed, typically by selecting a relevant time window and computing the mean force [7,19–21,38–41]. This adds processing time and may require additional steps for integration into standard clinical procedures.

To date, several studies have investigated operator dependency in HHD measurements [7,19–21,38–41]. Moreover, no study has yet implemented a serial configuration that enables a direct comparison between the outputs of a handheld dynamometer and those of a load cell, considered the reference standard for force measurement, to evaluate their

https://doi.org/10.3390/instruments10010002

relative agreement. In addition, the impact of different signal processing strategies applied to load cell data on the final strength value has not been systematically investigated, despite its potential influence on measurement interpretation and clinical usability.

This study proposes the design and implementation of a custom Computer-Aided Design-modeled (CAD-modeled) support system that enables the rigid, serial alignment of a commercial hand-held dynamometer and a load cell. The configuration ensures consistent, standardized measurements while reducing operator-related variability in a setup that closely reflects clinical conditions. Additionally, the study examines the agreement between the two measurement devices (i.e., the proposed system and the HHD) and assesses the impact of various signal selection approaches on the force values obtained from load cell recordings.

2. Materials and Methods

All experiments were conducted at the Lab of Motion Analysis of the Fondazione Policlinico Universitario Campus Bio-Medico. A Lafayette HHD (Model 01165A, Lafayette Instrument Company, Lafayette, IN, USA) and a load cell (AMTI MC3A, Advanced Mechanical Technology, Inc., Westford, MA, USA) were used to measure the strength of the shoulder muscles.

The Lafayette HHD is a handheld dynamometer with a measurement range up to approximately 1335 N and a resolution of 0.1 N. It uses a threshold-based acquisition system that triggers data recording once a predefined force is exceeded. The recording duration can be set between 1 s and 10 s. In this study, the recording duration was set to 5 s, corresponding to the instructed maximal voluntary contraction. The HHD operates through a threshold-based acquisition system, which automatically triggers data recording once a predefined force level is exceeded. The device outputs peak force, average force, and time to peak force, and includes internal memory for data storage. Manual stabilization is required during testing. The AMTI MC3A is a six-axis force/torque sensor with a vertical load capacity of 4400 N. It provides six outputs (Fx, Fy, Fz, Mx, My, Mz), corresponding to the three orthogonal force and moment components. Sensor sensitivity is 0.17 μV/(V·N), stiffness is 30.0×10^7 N/m, and crosstalk is below 2%. Due to reduced horizontal capacity, the z-axis was selected as the main direction for force acquisition. The load cell was fixed inside a modular vertical bracket using threaded inserts and through holes [42]. The bracket included a rotatable support that allowed the z-axis to be aligned with the direction of the applied force [42]. Additionally, the height of the support is adjustable from 0 to approximately 2 m, allowing measurements in upright, seated, or supine positions [42].

A dedicated housing was used to connect the HHD to the load cell in a serial configuration, designed to ensure precise alignment and reliable force transmission. The support structure was designed in CAD and fabricated with 3D printing. It consisted of two main components: a base to fix the load cell (Figure 1a) and a housing for the dynamometer (Figure 1b). Assembly was achieved using screws and bolts aligned with the load cell's mounting points (Figure 1c). To ensure accurate and consistent force transmission from the HHD to the load cell, the structure was designed to prevent force dispersion and contact between components. The support featured a hollow outer cylinder with a diameter slightly larger than the sensitive element of the load cell (37 mm) and a solid inner cylinder, which was slightly smaller and directly connected to the dynamometer housing. The two components fit together to form a stable assembly that replicates testing conditions for any load cell orientation.

https://doi.org/10.3390/instruments10010002

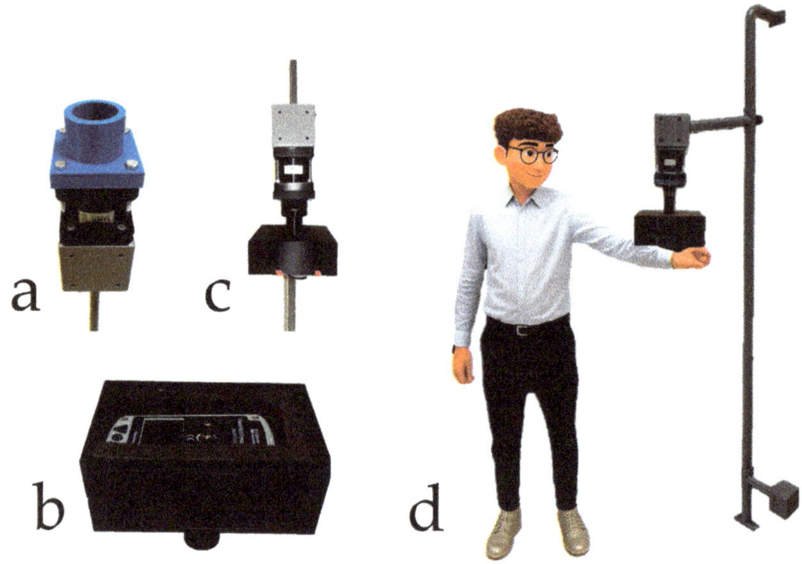

Figure 1. (**a**) A 3D-printed modular support securely attaches the load cell with screws and bolts. (**b**) A custom casing stabilizes the HHD. (**c**) The final configuration integrates the load cell, HHD, and support structure. (**d**) Example of a healthy participant performing one task from the experimental protocol: the participant was instructed to exert maximal muscle force against the two devices, arranged in a serial configuration.

To validate the structural accuracy of the system, six static weights (range: 9.81–98.10 N) were applied using cast-iron discs placed on a padded support above the HHD. This range was chosen as it reflects values reported in previous experiments and is consistent with data from the literature on patients with musculoskeletal shoulder disorders [14,30,34,43]. Measurements were performed using both devices in series, with 20 repetitions per weight, applied sequentially in increasing order. Both instruments were zeroed before each repetition. After acquisition, the force signals from the load cell were preprocessed to extract the relevant portion of the recording. From each repetition, the mean force value and the corresponding standard deviation (SD) were computed to assess the consistency of the measurements. To assess the agreement between the HHD and the load cell, the Root Mean Square Error (RMSE) was computed for each trial. The RMSE represents the square root of the average squared differences between the force values recorded by the two measurement systems [44]:

$$RMSE = \sqrt{\frac{1}{n}\sum_{i=1}^{n}\left(Force_{HHD_i} - Force_{LoadCell_i}\right)^2}$$

in which $\text{Force}_{\text{HHD}_i}$ represents the force measured by the HHD, and $\text{Force}_{\text{LoadCell}_i}$ refers to the force measured by the load cell, with n indicating the number of repetitions, and i representing the i-th repetition. This metric is particularly useful for capturing both the magnitude and variability in the error, as it increases proportionally with the presence of larger discrepancies in the data [44,45].

For human testing, a total of 5 healthy volunteers (mean age 26 years; 2 males and 3 females; 4 right-handed and 1 left-handed) were enrolled at the Fondazione Policlinico Universitario Campus Bio-Medico (FPUCBM). None of the participants had experienced shoulder pain in the past six months or had a history of shoulder surgery. The study was approved by the local Ethics Committee Lazio Area 2 (protocol code: 13.25 CET2 cbm), and

https://doi.org/10.3390/instruments10010002

informed consent was obtained from all participants prior to data collection. Participants were instructed to exert maximum muscle force against the instruments arranged in a serial configuration during the execution of two clinically relevant movements: arm elevation in the scapular plane (also known as scaption) and elbow flexion [19,41,46,47]. For the first movement, the participant stood with feet shoulder-width apart, arm abducted to 90° in the scapular plane, wrist pronated, elbow extended, and the HHD strap secured around the wrist, following the Constant Murley Score protocol [47]. The second exercise involved participants performing elbow flexion at 90 degrees in a standing position, with the forearm vertical and supinated, the upper arm stabilized against the trunk, and the HHD strap secured around the mid-forearm, aligned with the brachialis insertion, to ensure consistent force application. In clinical evaluation of shoulder muscle strength, elbow flexion is often employed to isolate and measure the strength of the biceps brachii, which is considered one of the major muscles contributing to shoulder stability and function. This movement allows the biceps to be tested in relative isolation and provides clinically relevant information on its role within the shoulder complex [19,41,46]. The test involved exerting maximal force against the system for 5 s while maintaining a steady and continuous effort throughout. Each participant completed five trials per arm to ensure consistency and reliability of the force measurements. A brief rest period (about 3 s) was provided between consecutive trials to maintain consistency in the testing procedure. Standardized instructions were provided during each trial, accompanied by clear cues that signaled the start and end of the test.

To synchronize the two systems, the HHD emitted an acoustic signal at both the start and the end of each trial. The load cell recording was initiated immediately after the start signal and terminated by the operator when the stop signal was emitted, using these acoustic markers as a common temporal reference.

The HHD data's mean and peak values were saved as a .csv file, which was subsequently imported into MATLAB® (R2024b, The MathWorks Inc., Natick, MA, USA) for further analysis. Load cell data were acquired using Qualisys Track Manager and exported as .mat files for further preprocessing in MATLAB®. A fourth-order Butterworth lowpass filter was applied to these data, and then the mean value was extracted for each muscle contraction signal. To assess the agreement between the two force-sensing systems and investigate how different signal selection methods influence the resulting force values, two distinct strategies were applied to identify the relevant portion of the load cell signal for each shoulder muscle strength curve. For both methods, the average force was computed over the selected segment. In the first method (rising phase), a 5 s window was extracted starting from the synchronization point with the HHD, which marked the onset of force application. This window captured the initial part of the contraction, including the ascending phase of the force curve, and ended at the time corresponding to the stop signal of the HHD. In the second method (plateau phase), a threshold-based approach was employed to isolate the more stable portion of the contraction [39,40]. Specifically, the rising phase of the signal was considered to end when the force reached 90% of its peak value [39,40]. From this point forward, all subsequent samples until the end of the trial (as defined by the stop signal from the HHD) were included in the analysis. This method was chosen to isolate the portion of the contraction most representative of maximal voluntary effort, minimizing the influence of initial acceleration and end-phase fatigue [48]. An illustrative example of the two signal selection methods was shown in Figure 2, where the rising and plateau phases are highlighted on a representative force signal.

https://doi.org/10.3390/instruments10010002

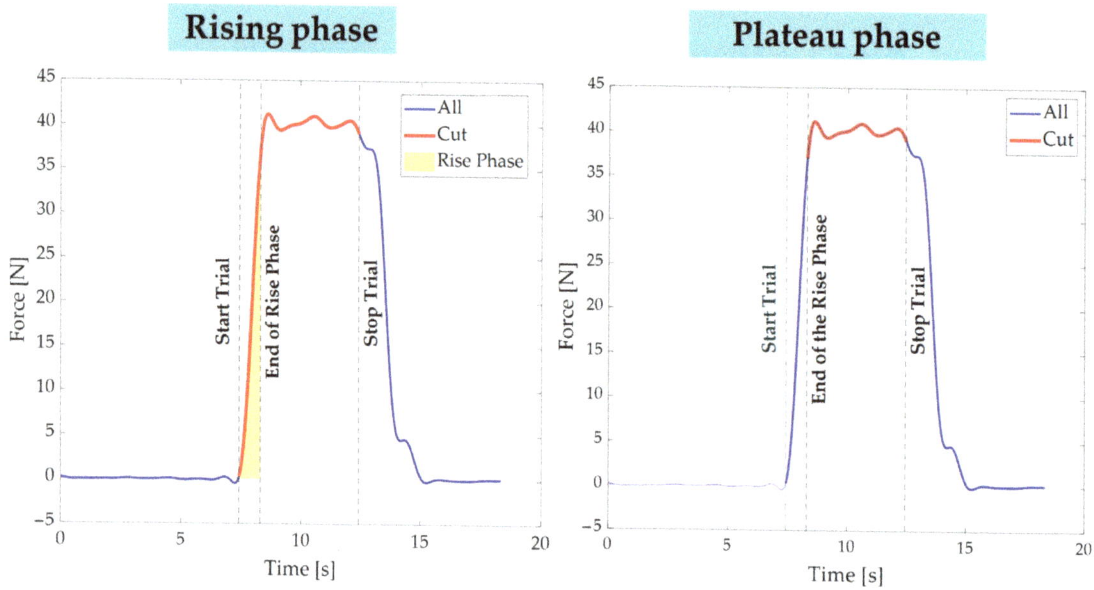

Figure 2. An example of a signal extracted using the load cell. The entire signal is shown in blue, while the portion used for calculating the mean muscle strength is highlighted in red. The left panel illustrates the Rising phase method, and the right panel shows the Plateau phase method.

To evaluate the agreement between the measurements obtained from the HHD and the load cell, a Bland–Altman analysis was performed. This method enabled a quantitative assessment of the agreement between the two instruments by calculating the mean of difference (MOD) and the limits of agreement (LOA) between paired force values [26,49–51]. MOD represents the average discrepancy between the two instruments, while LOA defines the interval within which approximately 95% of the data points are expected to fall [26,49–51]. These metrics were calculated as follows [49]:

$$MOD = \frac{1}{n}\sum_{i=1}^{n}\left(MeanStrength_{HHD_i} - MeanStrength_{LoadCell_i}\right)$$

$$LOA = MOD \pm 1.96\ SD$$

where SD is the standard deviation where SD is the standard deviation of the differences between the paired force values recorded by the two instruments. In addition, the RMSE was computed as an additional quantitative measure of agreement, following the same procedure adopted in the system validation with static weights, and calculated for each trial based on the point-by-point differences between the measurements obtained from the two force-measuring devices. Both analyses were conducted separately for each signal processing method (rising phase and plateau phase).

3. Results

All data acquired using the HHD and the load cell are presented in Table 1, which reports the mean force values and standard deviations for each static weight condition, along with the corresponding RMSE between the two devices. The load cell consistently demonstrated stable measurements across all applied loads. At 9.81 N, it recorded a mean value of 10.40 N (SD = 0.03 N), while at 98.10 N, it measured 98.61 N (SD = 0.25 N), maintaining a low variability throughout the tested range. These results reflect high precision,

https://doi.org/10.3390/instruments10010002

with standard deviations remaining well below 1% of the applied force. In comparison, the HHD exhibited increasing variability as the weight increased. For lighter loads, such as 9.81 N, the device reported 10.45 N (SD = 0.09 N), with a low RMSE of 0.55 N. Similar performance was observed at 19.62 N and 29.43 N (RMSE = 0.41 N and 0.46 N, respectively). However, for higher loads, dispersion around the mean became more pronounced. At 49.05 N, the HHD recorded 48.83 N (SD = 0.31 N), and at 98.10 N, it measured 93.38 N (SD = 0.60 N), with RMSE values increasing to 1.29 N and 5.06 N, respectively.

Table 1. Mean force values and standard deviations (SD) recorded by the HHD and the load cell across six known static weight conditions. The corresponding Root Mean Square Error (RMSE) quantifies the measurement discrepancy between the two devices for each load.

Known Weights [N]	HHD Mean Value (SD) [N]	Load Cell Mean Value (SD) [N]	RMSE [N]
9.81	10.45 (0.09)	10.40 (0.03)	0.55
19.62	19.76 (0.04)	19.79 (0.02)	0.41
29.43	30.04 (0.46)	30.29 (0.11)	0.46
39.24	38.63 (0.43)	39.37 (0.09)	0.40
49.05	48.83 (0.31)	50.08 (0.08)	1.29
98.10	93.38 (0.60)	98.81 (0.25)	5.06

Regarding the experimental tests conducted on healthy partecipants, the protocol was completed under controlled laboratory conditions, enabling the evaluation of measurement agreement and signal processing approaches in the absence of confounding clinical variability. A total of 50 paired measurements were collected and analyzed between the HHD and the load cell across two tasks (scaption and elbow flexion), both limbs, and five repetitions per condition. Mean force values and SD measured by the HHD and the load cell for both tasks were summarized in Tables 2 and 3. For each participant, these values represent the average of five repeated trials per condition.

Table 2. Mean force values and standard deviations (SD) recorded by the HHD and the load cell across healthy participants during the scaption.

Partecipant IDs	Arm Side	HHD Mean Value (SD) [N]	Rising Phase—Load Cell Mean Value (SD) [N]	Plateau Phase—Load Cell Mean Value (SD) [N]
1	Right	39.80 (7.07)	27.11 (3.95)	28.27 (4.42)
	Left	44.94 (7.53)	40.96 (8.59)	45.58 (19.94)
2	Right	30.36 (3.61)	20.55 (6.36)	24.10 (7.99)
	Left	31.36 (4.26)	22.60 (4.76)	26.30 (5.92)
3	Right	32.98 (3.28)	44.53 (9.82)	47.56 (10.95)
	Left	33.78 (10.96)	26.86 (3.74)	28.34 (4.27)
4	Right	19.80 (1.44)	11.60 (2.70)	12.11 (2.93)
	Left	19.76 (2.34)	12.25 (1.95)	13.15 (2.26)
5	Right	25.18 (1.31)	20.91 (3.51)	24.12 (4.99)
	Left	23.82 (1.67)	17.14 (1.71)	19.21 (2.33)

The Bland–Altman analysis results are presented in Figure 3, which were performed separately for the two signal selection techniques from the load cell: the rising phase and the plateau phase. Each subplot in Figure 3 reports the mean difference and limits of agreement for the respective task and processing approach. In the Scaption task, during the rising phase of the force signal, the MOD between the HHD and the load cell was 5.73 N. The LOAs were −12.94 N and +24.40 N, resulting in a total width of 37.34 N. This indicates a relevant difference between the two instruments during the ascending portion of the force curve, with a tendency of the HHD to slightly overestimate values compared to the

https://doi.org/10.3390/instruments10010002

load cell. In the plateau phase, the MOD decreased to 3.30 N, and the LOAs were −16.86 N and +23.46 N, with a total span of 40.32 N. Similarly, in the Elbow Flexion task, the rising phase showed a MOD of 2.68 N, with LOA of −21.95 N and +27.31 N, corresponding to a total range of 49.26 N. During the plateau phase, the MOD further decreased to −3.66 N, with LOA of −31.88 N and +24.55 N (range = 56.43 N). Despite the relatively low values of the MOD, the agreement between the devices remained limited due to the broad limits of agreement.

Table 3. Mean force values and standard deviations (SD) recorded by the HHD and the load cell across healthy participants during the elbow flexion task.

Partecipant IDs	Arm Side	HHD Mean Value (SD) [N]	Rising Phase—Load Cell Mean Value (SD) [N]	Plateau Phase—Load Cell Mean Value (SD) [N]
1	Right Left	81.66 (8.83) 82.34 (2.08)	73.46 (9.47) 78.58 (11.32)	80.26 (10.81) 89.78 (14.37)
2	Right Left	50.42 (6.48) 62.02 (5.53)	57.01 (6.87) 55.00 (10.88)	69.33 (9.27) 70.25 (11.71)
3	Right Left	43.24 (9.76) 46.32 (5.60)	51.54 (12.15) 54.78 (12.65)	54.42 (12.85) 57.55 (13.13)
4	Right Left	31.26 (1.92) 22.94 (4.40)	17.12 (3.03) 21.60 (3.57)	18.36 (3.39) 23.37 (4.21)
5	Right Left	43.12 (11.80) 37.06 (2.89)	28.61 (2.41) 35.90 (11.92)	32.27 (2.03) 41.42 (13.21)

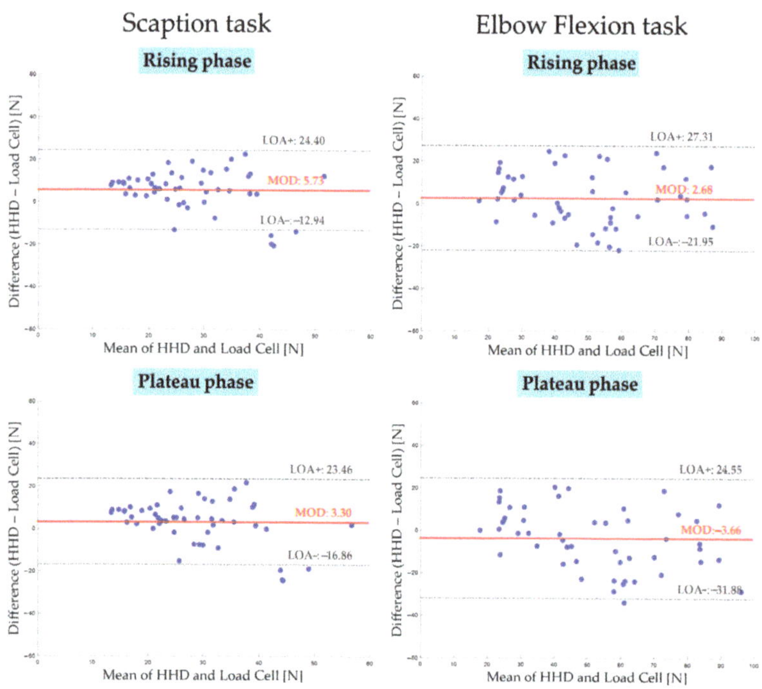

Figure 3. Bland–Altman plot comparing shoulder muscle strength measurements obtained using the two devices and both techniques for extracting the load cell mean value. The mean of differences (MOD) is represented by the solid red line, while the upper and lower limits of agreement (LOA) are indicated by the dashed grey lines.

https://doi.org/10.3390/instruments10010002

In addition to the Bland–Altman analysis, RMSE values were calculated to further quantify the magnitude of disagreement. For the Scaption task, RMSE was 11.03 N during the rising phase and 10.70 N during the plateau phase. In the Elbow Flexion task, RMSE was 12.72 N in the rising phase and increased to 14.71 N in the plateau phase.

4. Discussion and Conclusions

This study directly addressed the current lack of evidence on the methodological comparison between HHD measurements and those of a reference load cell by implementing, for the first time, a rigid serial configuration between a commercial hand-held dynamometer and a load cell. This design enabled both systems to be tested under identical loading conditions, thereby minimizing operator-related variability and allowing for a direct assessment of their relative accuracy and reliability. In addition, the work systematically examined how different load cell signal processing strategies affect the final measured force, providing insight into their impact on data interpretation and potential clinical applicability.

The static validation phase provided important insights into the performance characteristics of both measurement systems under controlled loading conditions. In the lower load conditions, both instruments slightly overestimated the applied force, which may be attributed to baseline signal noise and the high sensitivity of the transducers at low force levels. Conversely, in the higher load conditions, the HHD tended to underestimate the applied forces, likely due to minor mechanical deformation and limited stabilization of the handheld system, whereas the load cell continued to slightly overestimate, reflecting its higher stiffness and precision in load detection. In particular, the results showed that the load cell maintained standard deviations well below 1% of the applied load, confirming its suitability as a reference instrument for biomechanical force assessment in both clinical and research settings. Conversely, the HHD showed increasing dispersion at higher loads, with RMSE values rising sharply from 0.55 N at 9.81 N to 5.06 N at 98.10 N. This pattern reflects established evidence showing that portable dynamometry is more susceptible to measurement variability under high-force conditions, where tester stabilization and device handling can influence accuracy [16–18,22–25,30–33,42].

The study was conducted in healthy participants, and the experimental setup was designed to ensure consistent and comparable conditions for the methodological evaluation of the two measurement systems. During human testing, both systems reported mean force values that were relatively low compared with previously published normative data, which can be attributed to the controlled testing conditions, standardized posture, and short rest intervals intentionally adopted to enhance measurement reproducibility and minimize operator-related variability rather than to elicit maximal performance [52].

The variability of the HHD measurements also increased with higher force levels, mirroring the trend observed in the static weight validation. This finding is consistent with previous literature, which has reported that portable dynamometry tends to show increased dispersion at higher force outputs, particularly when examiner stabilization becomes a limiting factor [16–18,22–25,30–33,42]. Bland–Altman analysis showed that selecting the plateau phase generally reduced the MOD between devices in both scaption and elbow flexion tasks. In scaption, the MOD decreased from 5.73 N in the rising phase to 3.30 N in the plateau phase, while in elbow flexion it shifted from 2.68 N to −3.66 N. Although this change did not represent a true reduction, it indicated a shift in the direction of the bias, with the HHD tending to underestimate force values compared to the load cell. However, these differences were not consistently associated with narrower LOA or lower RMSE. For example, in elbow flexion, the LOA widened from 49.26 N to 56.43 N, and RMSE increased from 12.72 N to 14.71 N, corresponding to approximately 10–12%

https://doi.org/10.3390/instruments10010002

of the mean force values recorded in this task. Although these variations are unlikely to be clinically meaningful, they demonstrate that changes in signal selection criteria can affect the numerical comparison between devices. Likely contributors include differences in sensor design, sampling characteristics, and internal processing algorithms, as well as mechanical factors such as minor misalignments, variable fixation stability, and different dynamic responses to rapid force changes [53]. While excluding the initial ramp-up phase can reduce certain systematic biases, it may also eliminate essential information on contraction onset and early force development, periods that reflect fundamental aspects of neuromuscular function [54,55]. Its omission does not necessarily reduce variability and may compromise the physiological validity of the measurement [49]. However, it should be noted that the calculation of the mean force also depends on the internal algorithm of the HHD, which averages the sampled values over the entire test duration. This methodological difference may contribute to the discrepancies observed between devices.

Scaption consistently exhibited a narrower LOA than elbow flexion across both signal selection methods. This difference is likely explained by the biomechanical characteristics of the tasks [56]. In scaption, the peak forces were generally lower than those generated during elbow flexion, allowing the limb to be maintained in a more stable and controlled position throughout the contraction, as the reduced load minimized compensatory movements and instability [17,24,32–34,56]. This stability limits the occurrence of compensatory movements, reduces joint translation, and minimizes mechanical artifacts transmitted to the measuring devices [17,24,32–34,56]. In contrast, elbow flexion tasks typically involve higher peak forces and require the partecipant to counteract greater torque at the joint [17,24,32–34]. These higher demands place increased emphasis on both partecipant stabilization and precise alignment of the measurement apparatus [17,57]. Even in a rigid serial configuration, small deviations in posture, for example, may alter load transmission, leading to greater variability in recorded values. Furthermore, the elbow flexion posture may allow for subtle adjustments in trunk or shoulder position, although minor, which can affect force measurements and widen the LOA [58].

The serial configuration proposed in this study enabled a direct and controlled comparison between two measurement systems for assessing shoulder strength. The load cell consistently delivered stable and precise measurements across all conditions, whereas the HHD exhibited increasing variability as force levels rose. Selecting the plateau phase of the load cell signal reduced the MOD between devices; however, this adjustment did not consistently lead to narrower LOA or lower RMSE across conditions. Taken together, the findings suggest that the load cell would maintain high stability and precision in both tests with known weights and healthy participants. In contrast, the HHD displayed greater variability, which became more pronounced at higher applied forces. The method used to select the signal segment for analysis also influenced the results: plateau-phase analysis generally reduced MOD, yet this adjustment did not consistently improve LOA or RMSE. Moreover, the comparison between the two signal processing approaches aimed to highlight that methodological choices, such as whether to include or exclude the rising phase, can influence not only the numerical results but also the interpretation of force generation in a clinical setting. Understanding how these variations affect the agreement between devices is important for developing standardized analysis procedures and ensuring more consistent and comparable strength assessments in both research and clinical applications.

Differences between devices must be considered when interpreting absolute force values, particularly in clinical contexts, where even minor inaccuracies in strength estimation may influence the evaluation of functional recovery and subsequent therapeutic decisions. These findings underscore both the potential and the limitations of portable dynamometry in clinical practice, highlighting its practicality for routine strength assessment while also

https://doi.org/10.3390/instruments10010002

emphasizing its susceptibility to variability under higher-load conditions or when stabilization cannot be fully controlled, thereby underscoring the need for standardized and operator-independent clinical protocols supported by complementary reference validation. While well-suited for low- to moderate-force measurements, the accuracy of HHDs declines under high-load conditions, highlighting the importance of reference validation and, where feasible, the integration of objective signal processing techniques.

The proposed configuration establishes a controlled and repeatable approach for assessing shoulder muscle strength measurement using commercial devices, supporting the development of evidence-based standards for research and clinical applications. In fact, this setup provides objective data that may contribute to the standardization of measurement protocols, the definition of normative values, and the validation of portable devices for use in both laboratory and clinical settings. This study has some limitations. The sample was relatively small and homogeneous, comprising healthy young volunteers, which may limit the generalizability of the findings. Moreover, all tests were conducted under controlled laboratory conditions, and accuracy was not directly assessed in clinical populations. Additionally, although the Bland–Altman analysis assumes independent observations, in the present repeated-measures design, each measurement was treated as an independent data point due to the small sample size and the exploratory nature of this pilot study. Nonetheless, the controlled methodological design enabled a precise evaluation of the measurement systems while minimizing potential confounding factors and provided a preliminary understanding of the comparative performance between the two devices. Future work should assess the performance of this configuration in clinical populations and explore automated processing pipelines to streamline load cell signal analysis, thereby facilitating its integration into routine clinical practice. Reliable and reproducible strength measurement forms the basis for evidence-driven interventions, enabling clinicians to optimize patient function and quality of life.

Author Contributions: Conceptualization: C.A.; Methodology: A.C., U.G.L. and L.M.; Formal analysis: A.d.S. and P.D.; Investigation: M.M., C.A., E.S. and U.G.L.; Resources: A.C.; Data curation: A.C., L.M., E.S. and R.P.; Writing—Original Draft: C.A., A.C., L.M. and E.S.; Writing—Review & Editing: E.S., U.G.L. and A.d.S.; Visualization: R.P., P.D. and M.M.; Supervision: U.G.L. and E.S. All authors have read and agreed to the published version of the manuscript.

Funding: This research received no external funding.

Data Availability Statement: The original contributions presented in this study are included in the article. Further inquiries can be directed to the corresponding author.

Acknowledgments: This work has been supported by the Italian Ministry of Health in the framework of RICERCA FINALIZZATA 2021 (RF-2021-12372810).

Conflicts of Interest: The authors declare no conflicts of interest.

Abbreviations

The following abbreviations are used in this manuscript:

HHD	Hand-held Dynamometer
ST	Standard Deviation
RMSE	Root Mean Square Error
MOD	Mean of Difference
LOA	Limits of Agreement

https://doi.org/10.3390/instruments10010002

References

1. Matsuki, K.; Matsuki, K.O.; Mu, S.; Yamaguchi, S.; Ochiai, N.; Sasho, T.; Sugaya, H.; Toyone, T.; Wada, Y.; Takahashi, K.; et al. In vivo 3-dimensional analysis of scapular kinematics: Comparison of dominant and nondominant shoulders. *J. Shoulder Elb. Surg.* **2011**, *20*, 659–665. [CrossRef] [PubMed]
2. Maffulli, N.; Longo, U.G.; Berton, A.; Loppini, M.; Denaro, V. Biological factors in the pathogenesis of rotator cuff tears. *Sports Med. Arthrosc. Rev.* **2011**, *19*, 194–201. [CrossRef]
3. Rockwood, C.A. *The Shoulder*; Elsevier Health Sciences: Amsterdam, The Netherlands, 2009; Volume 1.
4. Longo, U.G.; Risi Ambrogioni, L.; Berton, A.; Candela, V.; Massaroni, C.; Carnevale, A.; Stelitano, G.; Schena, E.; Nazarian, A.; DeAngelis, J.; et al. Scapular Dyskinesis: From Basic Science to Ultimate Treatment. *Int. J. Environ. Res. Public Health* **2020**, *17*, 2974. [CrossRef]
5. Cools, A.M.; De Wilde, L.; Van Tongel, A.; Ceyssens, C.; Ryckewaert, R.; Cambier, D.C. Measuring shoulder external and internal rotation strength and range of motion: Comprehensive intra-rater and inter-rater reliability study of several testing protocols. *J. Shoulder Elb. Surg.* **2014**, *23*, 1454–1461. [CrossRef]
6. Mera, B.J. Current Perspectives on Rotator Cuff Disease. *Osteology* **2022**, *2*, 62–69. [CrossRef]
7. González-Rosalén, J.; Benítez-Martínez, J.C.; Medina-Mirapeix, F.; Cuerda-Del Pino, A.; Cervelló, A.; Martín-San Agustín, R. Intra- and Inter-Rater Reliability of Strength Measurements Using a Pull Hand-Held Dynamometer Fixed to the Examiner's Body and Comparison with Push Dynamometry. *Diagnostics* **2021**, *11*, 1230. [CrossRef]
8. Franceschi, F.; Longo, U.G.; Ruzzini, L.; Rizzello, G.; Denaro, V. Arthroscopic management of calcific tendinitis of the subscapularis tendon. *Knee Surg. Sports Traumatol. Arthrosc.* **2007**, *15*, 1482–1485. [CrossRef]
9. Maffulli, N.; Longo, U.G. Conservative management for tendinopathy: Is there enough scientific evidence? *Rheumatology* **2008**, *47*, 390–391. [CrossRef]
10. Constant, C.R.; Gerber, C.; Emery, R.J.; Søjbjerg, J.O.; Gohlke, F.; Boileau, P. A review of the Constant score: Modifications and guidelines for its use. *J. Shoulder Elb. Surg.* **2008**, *17*, 355–361. [CrossRef]
11. Vrotsou, K.; Ávila, M.; Machón, M.; Mateo-Abad, M.; Pardo, Y.; Garin, O.; Zaror, C.; González, N.; Escobar, A.; Cuéllar, R. Constant-Murley Score: Systematic review and standardized evaluation in different shoulder pathologies. *Qual. Life Res.* **2018**, *27*, 2217–2226. [CrossRef]
12. Booker, S.; Alfahad, N.; Scott, M.; Gooding, B.; Wallace, W.A. Use of scoring systems for assessing and reporting the outcome results from shoulder surgery and arthroplasty. *World J. Orthop.* **2015**, *6*, 244–251. [CrossRef] [PubMed]
13. Kirkley, A.; Griffin, S.; Dainty, K. Scoring systems for the functional assessment of the shoulder. *Arthroscopy* **2003**, *19*, 1109–1120. [CrossRef] [PubMed]
14. Ziegler, P.; Kühle, L.; Stöckle, U.; Wintermeyer, E.; Stollhof, L.E.; Ihle, C.; Bahrs, C. Evaluation of the Constant score: Which is the method to assess the objective strength? *BMC Musculoskelet. Disord.* **2019**, *20*, 403. [CrossRef]
15. Burrus, C.; Deriaz, O.; Luthi, F.; Konzelmann, M. Role of pain in measuring shoulder strength abduction and flexion with the Constant-Murley score. *Ann. Phys. Rehabil. Med.* **2017**, *60*, 258–262. [CrossRef]
16. Durfee, W.K.; Iaizzo, P. *Rehabilitation and Muscle Testing*; John Wiley and Sons: Hoboken, NJ, USA, 2006.
17. Wikholm, J.B.; Bohannon, R.W. Hand-held Dynamometer Measurements: Tester Strength Makes a Difference. *J. Orthop. Sports Phys. Ther.* **1991**, *13*, 191–198. [CrossRef]
18. Chamorro, C.; Arancibia, M.; Trigo, B.; Arias-Poblete, L.; Jerez-Mayorga, D. Absolute Reliability and Concurrent Validity of Hand-Held Dynamometry in Shoulder Rotator Strength Assessment: Systematic Review and Meta-Analysis. *Int. J. Environ. Res. Public Health* **2021**, *18*, 9293. [CrossRef]
19. Decleve, P.; Van Cant, J.; De Buck, E.; Van Doren, J.; Verkouille, J.; Cools, A.M. The Self-Assessment Corner for Shoulder Strength: Reliability, Validity, and Correlations With Upper Extremity Physical Performance Tests. *J. Athl. Train.* **2020**, *55*, 350–358. [CrossRef]
20. Hirano, M.; Katoh, M. Absolute reliability of shoulder joint horizontal adductor muscle strength measurements using a handheld dynamometer. *J. Phys. Ther. Sci.* **2015**, *27*, 2125–2127. [CrossRef]
21. Holt, K.L.; Raper, D.P.; Boettcher, C.E.; Waddington, G.S.; Drew, M.K. Hand-held dynamometry strength measures for internal and external rotation demonstrate superior reliability, lower minimal detectable change and higher correlation to isokinetic dynamometry than externally-fixed dynamometry of the shoulder. *Phys. Ther. Sport* **2016**, *21*, 75–81. [CrossRef]
22. Croci, E.; Born, P.; Eckers, F.; Nüesch, C.; Baumgartner, D.; Müller, A.M.; Mündermann, A. Test-retest reliability of isometric shoulder muscle strength during abduction and rotation tasks measured using the Biodex dynamometer. *J. Shoulder Elb. Surg.* **2023**, *32*, 2008–2016. [CrossRef] [PubMed]
23. Tanveer, F.; Arslan, S.A.; Darain, H.; Ahmad, A. Reliability of Hand-Held Dynamometer for assessing Isometric Lumbar Muscles Strength in Asymptomatic Healthy Population. *Pak. J. Med. Sci.* **2021**, *37*, 461–465. [CrossRef]
24. Byl, N.N.; Richards, S.; Asturias, J. Intrarater and interrater reliability of strength measurements of the biceps and deltoid using a hand held dynamometer. *J. Orthop. Sports Phys. Ther.* **1988**, *9*, 395–398. [CrossRef]

https://doi.org/10.3390/instruments10010002

25. Cadogan, A.; Laslett, M.; Hing, W.; McNair, P.; Williams, M. Reliability of a new hand-held dynamometer in measuring shoulder range of motion and strength. *Man. Ther.* **2011**, *16*, 97–101. [CrossRef]
26. Beshara, P.; Davidson, I.; Pelletier, M.; Walsh, W.R. The Intra- and Inter-Rater Reliability of a Variety of Testing Methods to Measure Shoulder Range of Motion, Hand-behind-Back and External Rotation Strength in Healthy Participants. *Int. J. Environ. Res. Public Health* **2022**, *19*, 14442. [CrossRef]
27. Stone, C.A.; Nolan, B.; Lawlor, P.G.; Kenny, R.A. Hand-held dynamometry: Tester strength is paramount, even in frail populations. *J. Rehabil. Med.* **2011**, *43*, 808–811. [CrossRef] [PubMed]
28. Wadsworth, C.T.; Krishnan, R.; Sear, M.; Harrold, J.; Nielsen, D.H. Intrarater reliability of manual muscle testing and hand-held dynametric muscle testing. *Phys. Ther.* **1987**, *67*, 1342–1347. [CrossRef] [PubMed]
29. Reed, R.L.; Den Hartog, R.; Yochum, K.; Pearlmutter, L.; Ruttinger, A.C.; Mooradian, A.D. A comparison of hand-held isometric strength measurement with isokinetic muscle strength measurement in the elderly. *J. Am. Geriatr. Soc.* **1993**, *41*, 53–56. [CrossRef] [PubMed]
30. Brinkmann, J.R. Comparison of a hand-held and fixed dynamometer in measuring strength of patients with neuromuscular disease. *J. Orthop. Sports Phys. Ther.* **1994**, *19*, 100–104. [CrossRef]
31. Bohannon, R.W. Intertester reliability of hand-held dynamometry: A concise summary of published research. *Percept. Mot. Ski.* **1999**, *88*, 899–902. [CrossRef]
32. McMahon, L.M.; Burdett, R.; Whitney, S.L. Effects of muscle group and placement site on reliability of hand-held dynamometry strength measurements. *J. Orthop. Sports Phys. Ther.* **1992**, *15*, 236–242. [CrossRef]
33. Lu, T.W.; Hsu, H.C.; Chang, L.Y.; Chen, H.L. Enhancing the examiner's resisting force improves the reliability of manual muscle strength measurements: Comparison of a new device with hand-held dynamometry. *J. Rehabil. Med.* **2007**, *39*, 679–684. [CrossRef]
34. Beshay, N.; Lam, P.H.; Murrell, G.A. Assessing the reliability of shoulder strength measurement: Hand-held versus fixed dynamometry. *Shoulder Elb.* **2011**, *3*, 244–251. [CrossRef]
35. Ruschel, C.; Haupenthal, A.; Jacomel, G.F.; Fontana, H.B.; Santos, D.P.; Scoz, R.D.; Roesler, H. Validity and reliability of an instrumented leg-extension machine for measuring isometric muscle strength of the knee extensors. *J. Sport Rehabil.* **2015**, *24*, 122. [CrossRef]
36. Macedo, M.C.; Souza, M.A.; Ferreira, K.R.; Campos, L.O.; Souza, I.S.O.; Barbosa, M.A.; Brito, C.J.; Intelangelo, L.; Barbosa, A.C. Validity and Test-Retest Reliability of a Novel Push Low-Cost Hand-Held Dynamometer for Knee Strength Assessment during Different Force Ranges. *Diagnostics* **2022**, *12*, 186. [CrossRef]
37. Machado Rodrigues, F.; Demeyer, H.; Hornikx, M.; Camillo, C.A.; Calik-Kutukcu, E.; Burtin, C.; Janssens, W.; Troosters, T.; Osadnik, C. Validity and reliability of strain gauge measurement of volitional quadriceps force in patients with COPD. *Chron. Respir. Dis.* **2017**, *14*, 289–297. [CrossRef]
38. Hong, J.-S.; Ko, J.-B.; Ju, M.-M.; Lee, B.-K.; Park, D.-S.; Lee, S.-H. The Reliability and Validity of an Isometric Knee Strength Measurement Device in Older Adult Individuals. *Sensors* **2025**, *25*, 2981. [CrossRef]
39. Choi, J.; Yeoh, W.L.; Loh, P.Y.; Muraki, S. Force and electromyography responses during isometric force release of different rates and step-down magnitudes. *Hum. Mov. Sci.* **2019**, *67*, 102516. [CrossRef]
40. Liu, J.; Qu, X.; Stone, M.H. Evaluation of force-time curve analysis methods in the isometric mid-thigh pull test. *Sports Biomech.* **2023**, *22*, 1381–1397. [CrossRef] [PubMed]
41. González-Rosalén, J.; Cuerda-Del Pino, A.; Sánchez-Barbadora, M.; Martín-San Agustín, R. Validity and reliability of the DiCI for the measurement of shoulder flexion and abduction strength in asymptomatic and symptomatic subjects. *PeerJ* **2021**, *9*, e11600. [CrossRef] [PubMed]
42. Antonacci, C.; Ghelli, G.; Carnevale, A.; Silvestri, S.; Schena, E.; Longo, U.G. A novel system integrating load cell for assessing shoulder muscle strength. In Proceedings of the 2024 IEEE International Workshop on Metrology for Industry 4.0 & IoT (MetroInd4. 0 & IoT), Firenze, Italy, 29–31 May 2024; pp. 591–596.
43. Bradley, H.; Pierpoint, L. Normative Values of Isometric Shoulder Strength Among Healthy Adults. *Int. J. Sports Phys. Ther.* **2023**, *18*, 977–988. [CrossRef] [PubMed]
44. Chai, T. Root Mean Square. In *Encyclopedia of Mathematical Geosciences*; Springer: Berlin/Heidelberg, Germany, 2022; pp. 1–3.
45. Chicco, D.; Warrens, M.J.; Jurman, G. The coefficient of determination R-squared is more informative than SMAPE, MAE, MAPE, MSE and RMSE in regression analysis evaluation. *Peerj Comput. Sci.* **2021**, *7*, e623. [CrossRef]
46. Wittstein, J.R.; Queen, R.; Abbey, A.; Toth, A.; Moorman, C.T. Isokinetic strength, endurance, and subjective outcomes after biceps tenotomy versus tenodesis: A postoperative study. *Am. J. Sports Med.* **2011**, *39*, 857–865. [CrossRef] [PubMed]
47. Constant, C.R.; Murley, A.H. A clinical method of functional assessment of the shoulder. *Clin. Orthop. Relat. Res.* **1987**, *214*, 160–164. [CrossRef]
48. Cools, A.M.; Vanderstukken, F.; Vereecken, F.; Duprez, M.; Heyman, K.; Goethals, N.; Johansson, F. Eccentric and isometric shoulder rotator cuff strength testing using a hand-held dynamometer: Reference values for overhead athletes. *Knee Surg. Sports Traumatol. Arthrosc.* **2016**, *24*, 3838–3847. [CrossRef]

https://doi.org/10.3390/instruments10010002

49. Bland, J.M.; Altman, D.G. Statistical methods for assessing agreement between two methods of clinical measurement. *Int. J. Nurs. Stud.* **2010**, *47*, 931–936. [CrossRef]
50. Awatani, T.; Morikita, I.; Shinohara, J.; Mori, S.; Nariai, M.; Tatsumi, Y.; Nagata, A.; Koshiba, H. Intra- and inter-rater reliability of isometric shoulder extensor and internal rotator strength measurements performed using a hand-held dynamometer. *J. Phys. Ther. Sci.* **2016**, *28*, 3054–3059. [CrossRef]
51. Ludbrook, J. Statistical techniques for comparing measurers and methods of measurement: A critical review. *Clin. Exp. Pharmacol. Physiol.* **2002**, *29*, 527–536. [CrossRef]
52. Andrews, A.W.; Thomas, M.W.; Bohannon, R.W. Normative values for isometric muscle force measurements obtained with hand-held dynamometers. *Phys. Ther.* **1996**, *76*, 248–259. [CrossRef] [PubMed]
53. Merry, K.; Napier, C.; Chung, V.; Hannigan, B.C.; MacPherson, M.; Menon, C.; Scott, A. The Validity and Reliability of Two Commercially Available Load Sensors for Clinical Strength Assessment. *Sensors* **2021**, *21*, 8399. [CrossRef]
54. Maffiuletti, N.A.; Aagaard, P.; Blazevich, A.J.; Folland, J.; Tillin, N.; Duchateau, J. Rate of force development: Physiological and methodological considerations. *Eur. J. Appl. Physiol.* **2016**, *116*, 1091–1116. [CrossRef]
55. Aagaard, P.; Simonsen, E.B.; Andersen, J.L.; Magnusson, P.; Dyhre-Poulsen, P. Increased rate of force development and neural drive of human skeletal muscle following resistance training. *J. Appl. Physiol.* **2002**, *93*, 1318–1326. [CrossRef] [PubMed]
56. Hoshikawa, K.; Yuri, T.; Giambini, H.; Kiyoshige, Y. Shoulder scaption is dependent on the behavior of the different partitions of the infraspinatus muscle. *Surg. Radiol. Anat.* **2021**, *43*, 653–659. [CrossRef]
57. Wang, Y.; Sun, Y.; Gao, L.; Ma, Z.; Yang, X. Development and performance verification of an isometric dynamometer for lower extremity. *Sci. Rep.* **2025**, *15*, 26145. [CrossRef] [PubMed]
58. Cavuoto, L.A.; Pajoutan, M.; Mehta, R.K. Reliability analyses and values of isometric shoulder flexion and trunk extension strengths stratified by body mass index. *PLoS ONE* **2019**, *14*, e0219090. [CrossRef] [PubMed]

https://doi.org/10.3390/instruments10010002

instruments

MDPI

Article

A Procedure for Performing Reproducibility Assessment of the Accuracy of Impact Area Classification for Structural Health Monitoring in Aerospace Structures †

Luciano Chiominto [1,*], Giulio D'Emilia [1], Antonella Gaspari [2], Emanuela Natale [1], Francesco Nicassio [3] and Gennaro Scarselli [4]

1 Department of Industrial and Information Engineering and Economics, University of L'Aquila, 67100 L'Aquila, Italy; giulio.demilia@univaq.it (G.D.); emanuela.natale@univaq.it (E.N.)
2 Department of Mechanics Mathematics and Management, Polytechnic of Bari, 70125 Bari, Italy; antonella.gaspari@poliba.it
3 Department of Engineering for Innovation, University of Salento, Via per Monteroni, 73100 Lecce, Italy; francesco.nicassio@unisalento.it
4 Department of Aeronautics and Astronautics, University of Southampton, Building 176, Boldrewood Innovation Campus, Burgess Road, Southampton SO16 7QF, UK; g.scarselli@soton.ac.uk
* Correspondence: luciano.chiominto@graduate.univaq.it
† This study is an extension of the authors' paper in Proceedings of the 2025 IEEE International Workshop on Metrology for Industry 4.0 & IoT (MetroInd4.0 & IoT), Castelldefels, Spain, 1–3 July 2025.

Abstract

The principal objective of this work is to develop an optimized procedure that guarantees the reproducibility of results across different applications and laboratories, facilitating potential field applications of methodologies for Structural Health Monitoring in aerospace structures. The focus is to accurately detect and localize impact areas on planar structures using in situ transducers and Machine Learning (ML) techniques. The research concentrates on an aluminum plate where impacts are generated by metal spheres of different masses dropped from a fixed height. The resulting Lamb waves are detected by PZT sensors glued on the surface. Various data processing and feature extraction algorithms are implemented and compared to extract the differences in Time of Flight (ΔToF). The obtained features are used for training ML classification models. Then, the influence of various parameters in signal acquisition and data processing are assessed along with the reproducibility of the results. For this reason, an interlaboratory comparison is conducted in which the trained models are applied to data collected under varying conditions. The experimental results show that the most influencing factors for impact area classification are the algorithm for ΔToF estimation, the number of training points used in ML models, the type of classification model, the distribution of the impact points on the component, and their balance in the classification area. This evidence suggests approaches for reducing both issues, therefore improving the reproducibility of results.

Keywords: measurement; validation; SHM; Lamb waves; piezoelectric sensors; machine learning; diagnostic; prognostic; classification; impact localization

1. Introduction

In the aerospace industry, ensuring structural integrity and early detection of defects is of paramount importance for several reasons. For example, structure monitoring improves safety, reduces unnecessary inspection and maintenance costs, and minimizes the weight of aircraft components during the design phase. To achieve these aims, non-destructive testing (NDT) techniques are commonly employed for structural assessments, often alongside theoretical models and simulations [1–6].

In recent decades, Structural Health Monitoring (SHM) has gained a lot of attention as its significance has become increasingly apparent. SHM involves a range of techniques that use sensor networks and sophisticated algorithms to provide real-time evaluations of structural reliability [7]. An SHM system is designed to detect structural damage in order to formulate a potential intervention plan. The damage identification process can be delineated into three primary stages: detection, diagnosis, and prognosis [8].

The first SHM phase is damage detection. This includes changes to the material and/or geometric properties of these structures which adversely affect the system's performance [9].

Diagnosis is an ongoing process throughout the structure's lifecycle. This stage aims to monitor its operational condition and identify any damage resulting from various factors such as material stress, collisions, or aging. This phase is further divided into passive and active diagnosis [10]. Passive diagnosis relies on sensors that capture any changes in the structure during normal use. In contrast, active diagnosis uses an actuator to apply stress, generating a response that is measured by receiving sensors, thus enabling assessment of the structure's condition.

The prognosis phase focuses on assessing the severity of any identified damage and estimating the remaining lifespan of the structure.

One of the key concepts of SHM is online monitoring, allowing all of the previously described phases to be carried out in embedded systems. These "smart" structures overcome several problems of traditional inspection methods. In fact, manual inspections, NDT, and model-based techniques are often labor-intensive, time-consuming, and may be inadequate for detecting hidden or developing damage [11]. This approach has several potential benefits:

- Facilitates condition-based maintenance.
- Prevents catastrophic damage.
- Reduces machine downtime.
- Eliminates human error.
- Monitors inaccessible components.

In the aerospace sector, the integrity and longevity of structures are significantly influenced by the frequency and intensity of impacts from orbiting objects, space debris, or, for aircraft, bird strikes [12].

One subset of SHM methods used for examining the impacts of external objects on a thin planar plate uses a passive diagnosis system in conjunction with the analysis of Lamb waves. Lamb waves represent an acoustic emission (AE) phenomenon and are a combination of longitudinal and transverse modes. The propagation properties of these waves depend on the excitation source and the structural geometry. Another important aspect of Lamb waves is their dispersive behavior. This means that propagation speed is frequency-dependent. A significant advantage of these waves is their ability to travel over long distances, thereby eliminating the need to access the structure using local detection methods [1,13].

To identify the impact damage, several studies have used Machine Learning (ML) techniques applied on AE signals. In [14], the authors developed an SHM system that identifies the source of an AE, allowing them to study the A0 mode of Lamb waves in a composite structure using PZT ($Pb[Zr_xTi_{1-x}]O_3$) sensors. The signals are analyzed by means of Continuous Wavelet Transform (CWT) and then the impact coordinates are determined by solving a set of non-linear equations using a combination of a local Newton

https://doi.org/10.3390/instruments10010006

iterative method and global unconstrained optimization. Other impact point localization algorithms are presented in the literature [15,16]. These methods do not use ML techniques and focus on the use of analytical techniques like FEM models combined with specific experimental setups. In all of these cases, the authors use closed-form mathematical solutions. Other approaches used for localizing impacts take advantage of ML techniques, as in [17]. Here, the authors used pre-trained stacked autoencoders in a two-step approach that first localizes AE sources and then characterizes them. Another method is presented in [18] and is based on the time reversal technique and ML approaches. This algorithm consists of two steps: The first involves a training process, where a baseline of structural responses from impact tests is computed. The second step assesses the impact location by utilizing the highest cross-correlation coefficient, derived from interpolating the impact response baseline. Other approaches are based on different models, as presented in [19]. In that study, automatic impact detection and localization use Random Forest and Stacked Autoencoder (SAE) techniques. A steel ball is dropped on a composite aircraft elevator to collect AE signals. The Random Forest algorithm is trained on 3600 AE samples with 15 features to identify impact zones, while the SAE utilizes both raw AE signals and their fast Fourier Transform (FFT) for impact localization. All of these studies demonstrate the potential of applying signal processing and ML techniques in SHM based on AEs with reference to specific cases and applications. Nevertheless, an important requirement to be discussed concerns the reproducibility of each approach, as experimental methods must be able to achieve accurate results even when used in different situations. Accurate bias and uncertainty analysis mitigates the effect of systematic and random errors. It is essential to quantitively estimate those effects to support interlaboratory comparisons, which are useful for assessing the effects of the such improvements [20]. This is more relevant when ML methods are used.

Based on the above considerations, this paper aims to assess the effects of many aspects of the procedure, from the measurement phase to data processing, on the final reproducibility of the data, including a comparison of results with measurements carried out in different laboratories. This paper is an extension of a preliminary work presented by the authors at the IEEE MetroInd 2025 conference [21]. The present study aims to evaluate the accuracy of different methods that use ML algorithms for identifying the area of impact of metal spheres of different masses. These are dropped on the surface of an aluminum plate in different locations, therefore generating elastic waves that propagate through the component and are measured by Piezo Wafer Active sensors (PWASs). The feature of interest for impact localization is the difference between the Times of Flight (ΔToFs) of two vibration signals. The content presented in [21] is expanded here, as we assess the influence of different parameters on the variability of the results and the possible systematic effects that influence the localization capabilities and variability of the trained models. In particular, the aspects examined are the number of impact points of each area of interest, its layout, and the strategy for generating and analyzing the impacts. With reference to this, the parameters evaluated are the type and material of spheres, the number of repeated impacts on the same point used for building the training dataset for the ML models, the algorithm used for ΔToF evaluation, and the frequency of acquisition of the acoustic signal. Moreover, different classification algorithms are compared for localizing the impacts areas. To assess the reproducibility of the methods and results, an interlaboratory comparison is performed where the trained models are used on data acquired in different test and laboratory conditions.

In Section 2, the test bench and the data acquisition and processing techniques are described, highlighting the parameters that influence the variability of results, which are theoretically and experimentally analyzed. In Section 3, the results of the parametric

https://doi.org/10.3390/instruments10010006

analysis are described and discussed to evaluate the most important parameters to be kept under control. Finally, an interlaboratory comparison is presented to assess the reproducibility of the experimental information.

The final section of this paper presents the conclusions.

2. Materials and Methods

2.1. Experimental Setup and Measurement Chain

The experimental setup centers on an aluminum alloy plate, which is a component of a complex structure, such as fuselage skin or wing surfaces, in a real-world context. The experimental setup used in this work is presented in Figure 1.

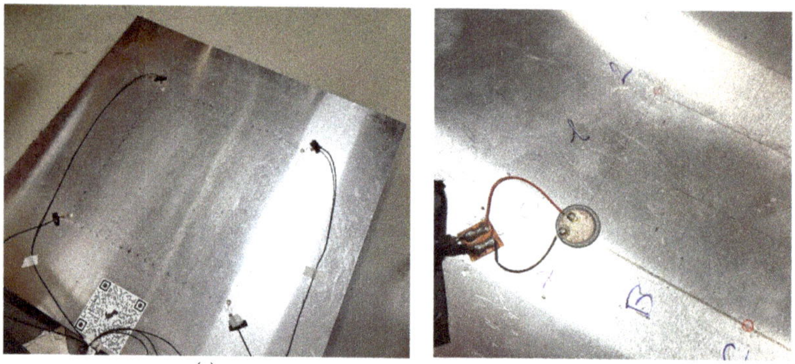

(a) (b)

Figure 1. Aluminum plate (**a**) under analysis and PZT sensor used to sense impacts (**b**).

As shown in Figure 1a, the aluminum alloy plate utilized in the experiment is representative of a portion of a panel used in the aerospace sector. It is 1000 mm in both length and width, with a thickness of 1.5 mm. It has an elastic modulus of 72 GPa, a Poisson's ratio of 0.33, and a density of 2700 kg/m^3. This plate is positioned on a foam rubber mattress that dampens its movements. This setup avoids unwanted vibrations and facilitates the detection of acoustic waves that propagate solely within the plate's thickness as Lamb waves.

As shown in Figure 1b, for detecting the impacts, four PZT ($Pb[Zr_xTi_{1-x}]O_3$) piezoelectric ceramic sensors are secured to the plate surface using bicomponent epoxy glue. The sensor model is PIC255 by Physik Instrumente (Physik Instrumente, Karlsruhe, Germany) [22], characterized by a diameter of 10 mm. The PZT sensors are positioned at the corners of a square with a side length of 500 mm. The sensors are placed 240 mm from the edges of the plate. This placement helps to ensure a cleaner initial impact signal. The reverberations due to the edges are delayed since the impact wave travels 240 mm + 240 mm before being registered by the sensor.

As depicted in Figure 2, a regular grid of 21 × 21 points defines the impact locations on the aluminum plate in the experimental tests, where the four positions (positions 1, 2, 3, and 4) at the vertex of the squared grid are occupied by the sensors. In total, 441 − 4 (sensor positions) = 437 impact points are realized on the component surface, at 25.0 mm from each other.

For generating the impacts, two spheres, namely S1 and S2, with different masses, are dropped on the plate surface. S1 is an 8.3 g stainless-steel sphere with a diameter of 12.6 mm, while S2 is a 3.6 g lead sphere with a diameter of 8.5 mm. As shown in Figure 3b, these are dropped from a fixed height of 330 mm using a tube for guiding the fall of the spheres at each impact point.

https://doi.org/10.3390/instruments10010006

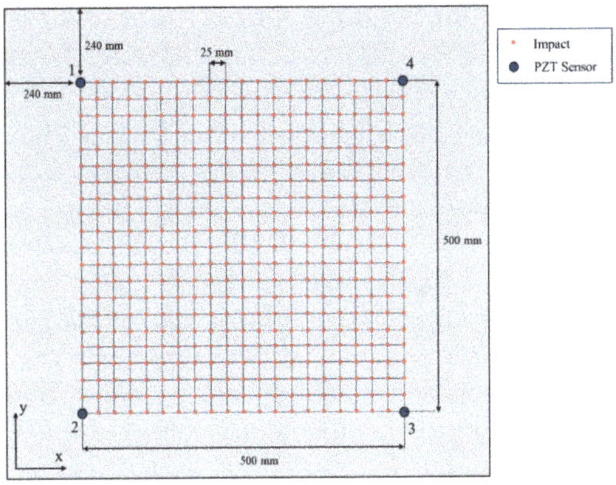

Figure 2. A scheme of the impact locations defined on the aluminum plate and placement of the PZT sensors.

(**a**) (**b**)

Figure 3. The spheres used for generating the impact on the plate surface (**a**) and a detailed image of the process (**b**).

The data from the four PZT sensors are acquired using a Siglent (Siglent, Shenzhen China) SDS824X HD 4-channel oscilloscope, connected to a PC using an Ethernet cable to download the data. The sampling frequency of the signals is 5 MHz, and the resolution is 12 bit. The vertical scale of the oscilloscope is set to 10.0 V/div, allowing signals in range $\pm$ 40.0 V. To start the sampling, the acquisition channels of the oscilloscope are triggered with a threshold equal to 1.0 V. The total recording length is set to 200 ms including a pre-trigger of 20 ms. In this manner, in each signal, the whole impact phenomenon is acquired. It represents the impact event itself and the subsequent vibrations on the plate until it returns to a steady state. Even though the above-mentioned reverberations are not the object of this study, some considerations are added in the following paragraphs.

The test campaign includes 437 impacts repeated two times using sphere S1 and 437 impacts repeated two times using sphere S2. In total, 1748 impact signals are acquired from each PZT sensor.

2.2. Data Processing

In thin aluminum plates, only the zero-modes of Lamb waves, specifically the symmetric (S0) and antisymmetric (A0) mode, can be activated, and consequently, the acoustic emission frequency spectrum is predominantly concentrated below 100 kHz [21].

Previous research by the authors revealed that the signal's arrival time, or Time of Flight (ToF), is primarily attributed to the A0 antisymmetric mode [10]. This is because

https://doi.org/10.3390/instruments10010006

the symmetric mode S0 exhibits negligible amplitudes within this frequency range. For this reason, the following analysis considers the Lamb waves in A0 mode at the specific frequency of 40 kHz.

Since the exact time instant of impact cannot be determined, the ToF is assessed from the start of acquisition. To determine the impact location, the key parameter is the difference in ToFs (ΔToF) recorded by the PZT sensors. The ΔToFs are estimated using Cross Correlation (CC), Continuous Wavelet Transform (CWT), and Short-Time Fourier Transform (STFT). All of the data processing algorithms are implemented in MATLAB 2023b.

Figure 4 shows the impact signals measured by two sensors, demonstrating the ΔToF concept when sensor 1 (blue) and sensor 3 (orange) are considered. In this case, for instance, the two vertical lines indicate the temporal instance when the signals start changing from a steady state. It can also be noted that a pre-trigger condition is set on the oscilloscope, as already stated in Section 2.1. Measuring the difference in Time of Flight at the very beginning of the signal was found to be the most repeatable way; setting a threshold caused higher variability, probably since the starting slope at the beginning is not uniform depending on position.

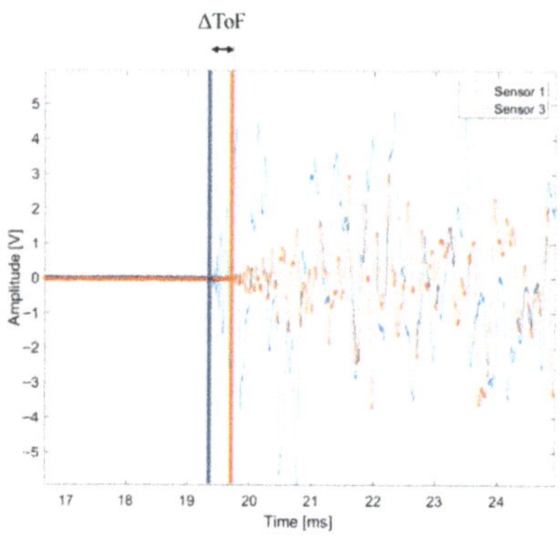

Figure 4. Example of ΔToF of impact measured by two sensors.

2.2.1. Cross Correlation

The oscilloscope signal is filtered using a bandpass Butterworth's filter of order 4. The selected cutoff frequencies are 35 kHz and 45 kHz, enabling us to analyze only the 40 kHz signal component. This is the frequency at which a 1.5 mm thick aluminum plate has the maximum A0 mode amplitude [10]. Then, using the MATLAB built-in function findchangepts() [23], the data index where the signal changes most significantly is identified. The option used in this function is "std", detecting changes in the standard deviation. The first instant of arrival of the wave at the four sensors is identified, and starting from this time point, considering the duration of the phenomenon, a time interval of 0.6 ms for all four acquired signals is selected for analysis. Considering an impact close to a sensor, 0.6 ms is the minimum time that it takes for the impact wave to travel from that point to the most distant sensor without introducing many reverberations in the first signal. The maximum distance found on the plate is 700 mm and the corresponding wave velocity is about 1135 m/s [10].

Successive waves are due to reflections and reverberations of the impact inside the plate. However, the selected signal length reduces unwanted vibrations. These signals are compared in pairs by calculating the CC coefficient [24]. The Cross Correlation coefficient peak corresponds to the associated delay between the two signals, which is the ΔToF. This procedure is repeated for all six sensor pairs.

https://doi.org/10.3390/instruments10010006

The flowchart of this procedure is represented in Figure 5.

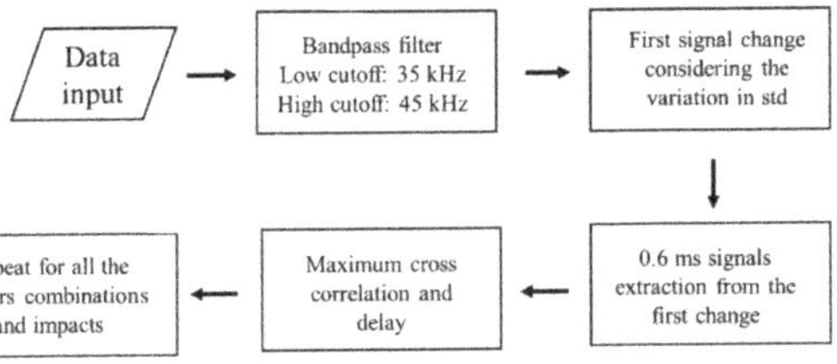

Figure 5. Flowchart of ΔToF estimation using CC.

2.2.2. Continuous Wavelet Transform

The impact signal for each sensor is analyzed using CWT to extract both time and frequency information. CWT uses a mother wavelet that is scaled and shifted to provide different window sizes for different frequencies. In this case, the CWT of the signal is computed in frequencies starting from 20 kHz to 60 kHz using Morse wavelets with a symmetry parameter equal to 3 and a time-bandwidth product equal to 60. This type of wavelet is considered useful in localizing discontinuities in signals [25]. To better represent the signal, the wavelet is computed using 16 voices per octave. The amplitude of the scale closer to the target frequency is considered, and its squared modulus is calculated. The impact instant is the time that corresponds to the first significant peak in the scalogram. The ΔToFs are the differences between the time instants identified for each sensor. All of these steps are summarized in Figure 6.

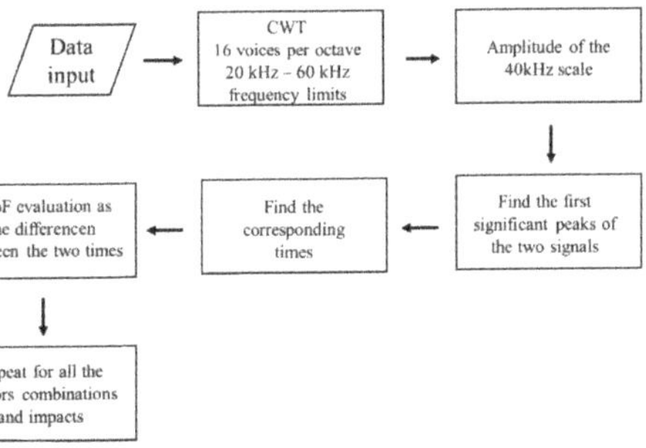

Figure 6. Flowchart of ΔToF estimation using CWT.

2.2.3. Short-Time Fourier Transform

Calculating STFT, the signal is analyzed both in time and frequency domains. In this case, the window length is 0.1 ms and the corresponding frequency resolution is equal to 10 kHz. In STFT, the window length is fixed for all frequencies. Selection of the window length must balance the necessity of frequency and time resolution. To reduce leakage, hamming windows are used with 90% overlapping. The ToF is calculated considering the amplitude of the signal corresponding to 40 kHz. Then, the local peaks of the energy at this frequency are identified using the built-in Matlab function findpeaks() [26]. This function identifies local maxima in a signal vector depending on the option 'MinPeakHeight', which is employed for refining the findings of the peaks. In this case, this option is activated to reduce the effect of initial noise on the measured signal. The first notable peak in the signal is recognized as the impact instant. Therefore, its time of occurrence indicates the ToF for each sensor, and the ΔToFs are determined considering all sensor pairs for each impact point.

https://doi.org/10.3390/instruments10010006

The variability in each ΔToF estimation is calculated with reference to each sensor pair considering four repeated impacts and assuming a uniform distribution [27].

The calculated ΔToFs considering all points and all repetitions represent the predictors for localizing impacts using Machine Learning approaches.

Figure 7 illustrates ΔToF estimation using the STFT methodology.

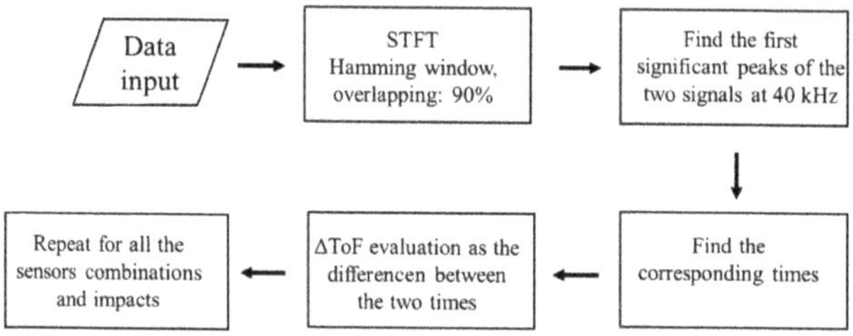

Figure 7. Flowchart of ΔToF estimation using STFT.

2.3. Machine Learning Models

This work focuses on using classification algorithms applied to the impact ΔToFs for identifying the regions of the plate where impacts occurred.

Since there are many possible classification models for this task and it is not possible to identify the best one a priori, multiple models are trained and compared in terms of classification accuracy.

2.3.1. Definition of Classification Zones

The definition of the plate zones is based on geometrical considerations to avoid biases in classification using ML algorithms. In this case, the impact point distribution is uniform on the surface; so, in each sub-area, there is the same number of points. This aspect is important for generating a training dataset that does not present class imbalances.

The aluminum plate surface is subdivided considering a matrix of 3 × 3 elements, resulting in 9 impact zones. Figure 8 shows the subdivision of components.

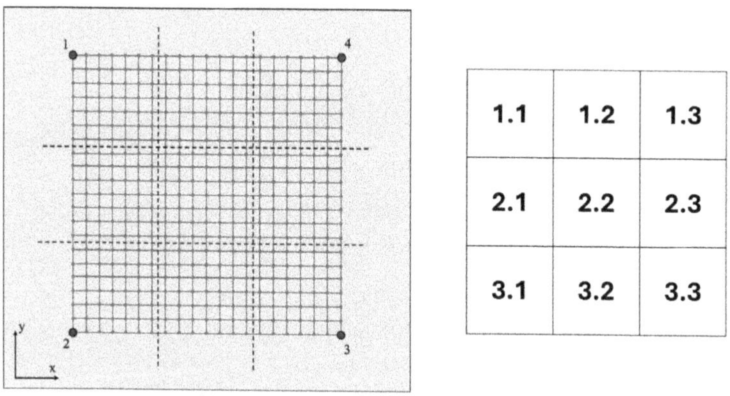

(**a**) (**b**)

Figure 8. Impact zones (**a**) and naming for each defined area (**b**). The red dots represent the impact points on the plate while the blue ones are the sensors. The black dot lines are the ideal subdivision of the plate surface.

The impact points are in a square measuring 150 × 150 mm. Using this subdivision, the number of impact points is balanced for each zone and equal to 49, or 48 for areas

https://doi.org/10.3390/instruments10010006

positioned at the edges, where a position is occupied by the PZT sensor. A representation of the zone nomenclature is reported in Figure 8b.

2.3.2. Effect of Impact Points Configuration

The points near the borders are the most critical. To evaluate their effect, a different point configuration is considered that excludes them. Each zone is redefined as shown in Figure 9.

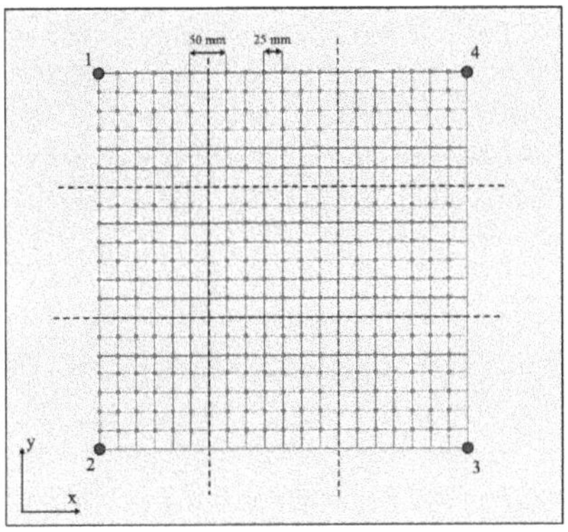

Figure 9. A scheme of the aluminum plate used for evaluating the effect of impact point layout. The red dots indicate the impact points on the plate, while the blue dots represent the sensors. The black dashed lines illustrate the subdivision of the plate's surface.

In this configuration, points that are located less than 25 mm from the border are not considered. Compared to the previous case, the dataset is 70% of the original one.

2.3.3. Effect of Number of Points

The effect of the amount of data used for training on classification performance is assessed by considering half impact points. Figure 10 represents the distribution of the impact points used in this analysis.

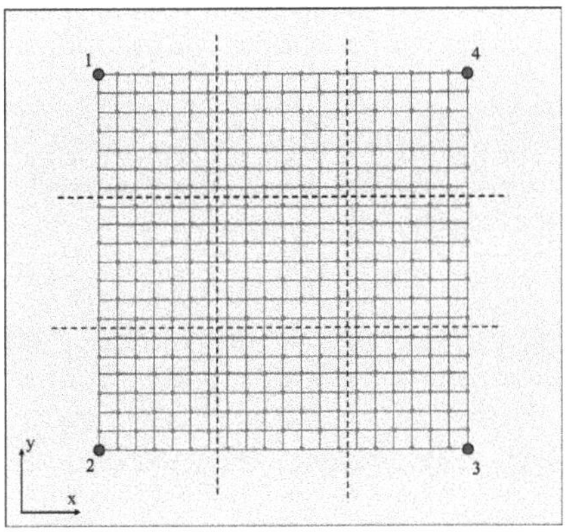

Figure 10. A representation of the impact point distribution for evaluating the effect of the amount of data used for training. The sensors are blue dots while the impacts are represented by the red dots. The dashed lines are the borders of each zone.

https://doi.org/10.3390/instruments10010006

With the impact points distributed as shown in Figure 10, along with data reduction, the minimum distance between two points is equal to ~35.4 mm, i.e., the diagonal of the square with dimensions of 25.0 mm.

2.3.4. Effect of Signal Sampling Frequency

Signal sampling frequency is important for several aspects of this analysis. First, it defines the maximum frequency that could be analyzed, according to the Nyquist theorem. Secondly, it influences the number of data points recorded and the further computational cost for ΔToF estimation. To evaluate the effect of the sampling frequency, the original 5.00 MHz signal is resampled to 1.00 MHz and 500 kHz. To perform signal resampling, the MATLAB resample() [28] function is used, allowing the input signal to be resampled with a pre-defined ratio between positive integers. This means that not only the input signal but also the constants 'p' and 'q' are provided; so, the function resamples the input sequence at p/q times the original sample rate.

2.3.5. Reproducibility of Results

To verify the robustness of the classification methods developed in this work with the aim of assessing the reproducibility of the results obtained, the impact signals acquired using a different experimental setup are analyzed. In particular, the same aluminum plate along with the PZT sensors is used for recording 167 impact signals. In this case, the data are sampled at 3.96 MHz using a different instrument [10] and the impacts are generated with bodies of different masses. To maintain the vibrational behavior of the plate, it is put on the same foam material.

The provided impact signals are then analyzed using the three developed methods and the corresponding ΔToFs are classified using the trained models.

Figure 11 shows a reference scheme of the general flowchart described in this paper. Regarding the ΔToF estimation methods, these are better described in Appendix A, where the pseudocodes are provided for each technique.

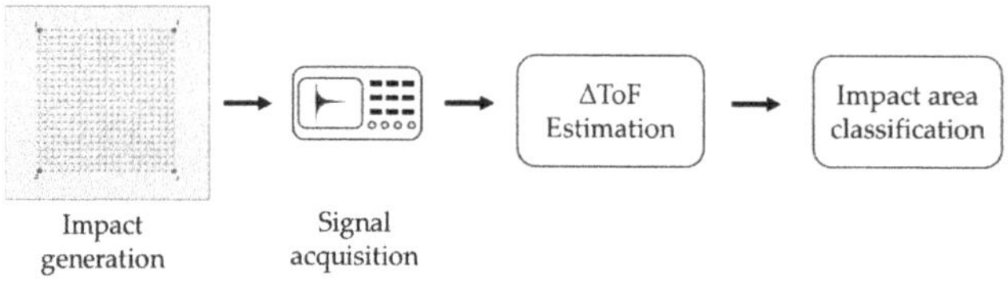

Figure 11. Scheme of processing chain for impact classification. The impacts depicted as red dots are generated on the aluminum plate. The signals are acquired using an oscilloscope and then are analyzed for estimating the ΔToFs and for classifying the impact area.

3. Results

Using the experimental setup outlined in the previous section, an example of the temporal signals acquired during a single impact is presented in Figure 12.

In this section, the results of the ΔToF estimation methods and the training of classification models are presented. As stated in Section 2, the influencing factors analyzed are as follows:

- ΔToF estimation method.
- Classification model.
- Layout of impact points.
- Sampling frequency of signals.

https://doi.org/10.3390/instruments10010006

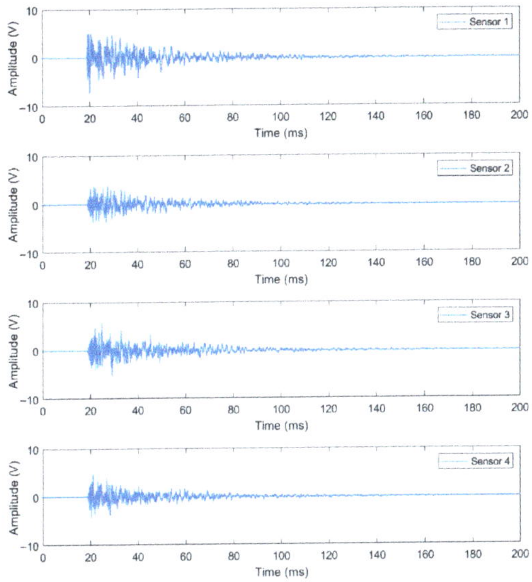

Figure 12. Temporal signal during impact using PZT sensors.

3.1. Differences Among ΔToF Estimation Methods

Figure 13 shows the estimations of the ΔToFs of a pair of sensors using the three different methods and considering both spheres for generating the impacts. The impact point coordinates as defined on the aluminum plate are represented on the X and Y axes, while the ΔToFs are reported on the Z axis.

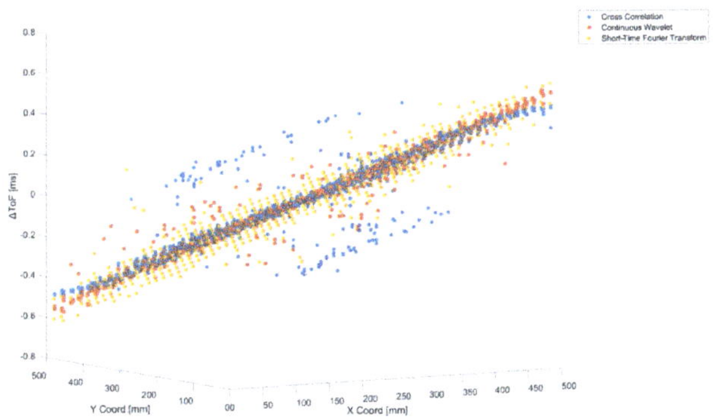

Figure 13. Resulting ΔToFs using all estimation methods.

In this case, the ΔToFs of sensor 1 and sensor 3 in a pair are reported, placed on the diagonal of the regular grid. Sensor 1 is the reference and has coordinates (0, 500) while sensor 3 is placed at (500, 0). Since the plate material is homogeneous with no discontinuities, the wave velocity inside it is constant. Consequently, the ΔToFs of the sensors placed on the diagonal are distributed on a plane, as can be seen in Figure 10.

CC is the method that presents more points that deviate more from the ideal behavior, clearly visible in Figure 10. This result is due to the method used for extracting the initial part of the signal. In some cases, in fact, the selected signal length also includes reverberations that affect the analysis. Figure 14 reports examples of different signals analyzed using the CC method. Figure 14a shows a signal where only the initial part

https://doi.org/10.3390/instruments10010006

of the impact is extracted, while Figure 14b represents a case where there are also some reverberations. It has to be noted that the represented signals are from two different impacts.

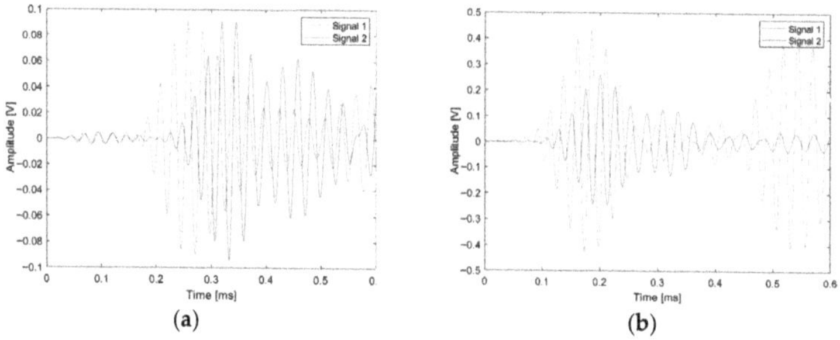

Figure 14. Examples of signals analyzed using CC. (**a**) only the initial impact is present. (**b**) some reverberations are present. Signal 1 and signal 2 refer to two different impacts.

Considering the ΔToFs estimated using STFT, the values are more diffused along the ideal fitting plane of the points shown in Figure 13. This is due to the temporal resolution linked to the selected window length.

For CWT, the points are closer and less diffused near the ideal plane due to the better temporal resolution of this method.

The variabilities of the ΔToFs are evaluated across four repeated impacts in all locations. In Figure 13, as an example, the ΔToF variability of sensor pair 1–3 is reported for each estimation method.

For CC in Figure 15a, ΔToF variability values are in the range from 0.0004 ms to 0.20 ms. The greater variability is acknowledged in the direction of the line connecting the two sensors, along the diagonal of the grid. The same behavior occurs considering the other pairs of sensors. This appears to be a consequence of the signal extraction method used for selecting which part to analyze, as described in Section 2.

The ΔToFs estimated using the CWT method have a minimum variability of 0.0002 ms and a maximum value of 0.0974 ms. The impacts with greater variability appear to be randomly distributed on the grid.

Finally, the minimum variability of STFT is equal to zero while the maximum value is 0.14 ms. Also, in this case, point variability distribution does not present a regular pattern.

The mean ΔToF variability of the different estimation methods is summarized in Table 1.

Table 1. Mean variability of different ΔToF estimation methods.

ΔToF Estimation Method	Mean Variability [ms]
CC	0.012
CWT	0.008
STFT	0.019

As reported in Table 1, CWT presents less variability compared to the other two approaches.

A repeatability test is performed, repeating 20 impacts on one location and considering the signals from sensor 1 and sensor 3. The ΔToFs are calculated using the three estimation methods and the repeatability is calculated considering the standard deviation. The repeatability of the ΔToFs estimated using CC is equal to 0.009 ms, while with CWT it is 0.002 ms and using STFT it is 0.030 ms. These results are satisfactory and prove that the

https://doi.org/10.3390/instruments10010006

handheld tube with careful positioning ensures the repeatability of the impacts within a maximum variability of 4–6%.

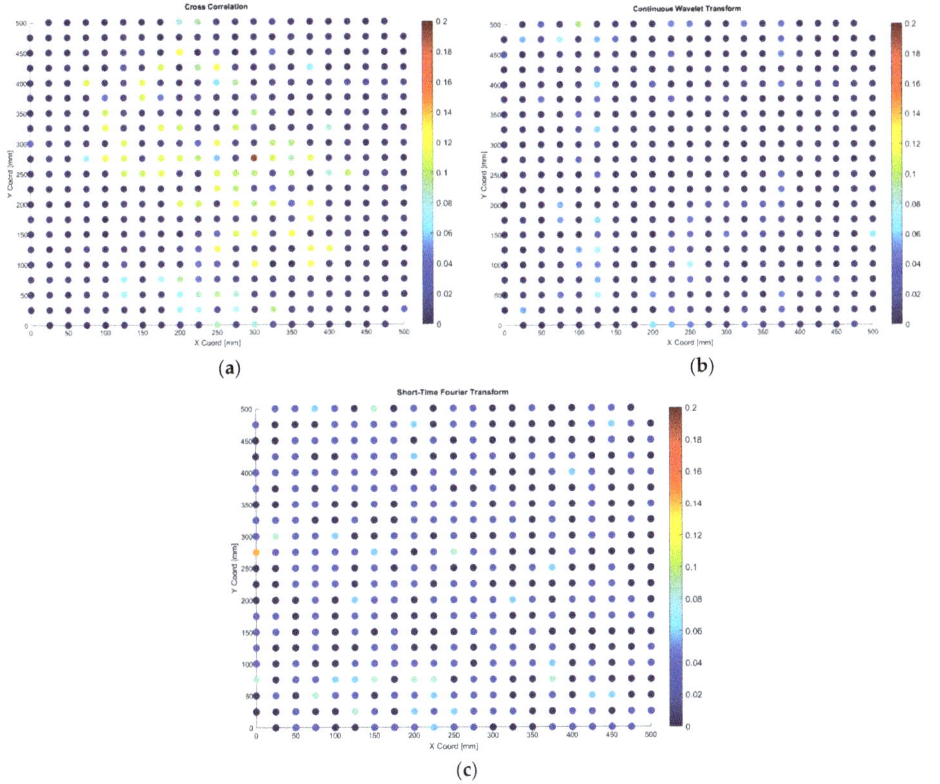

(**a**) (**b**) (**c**)

Figure 15. Variability of ΔToFs using Cross Correlation (**a**), Continuous Wavelet Transform (**b**), and Short-Time Fourier Transform (**c**).

3.2. Classification Models

As described in the previous section, the dataset is composed of 1748 impact signals, of which 80% are used for training while 20% are for model testing. To avoid overfitting, a k-fold cross-validation scheme is employed, partitioning the training dataset into five subfolders.

The features used as input of the classification models are the six ΔToFs from all sensor pairs, while the output is the impact zone.

A classification model is trained for each ΔToF estimation method, and different ML models are trained to find the best one. Testing accuracy is used as the first comparison metric of the different classifiers. Table 2 summarizes the performance of the trained classification models.

Table 2. Performance of trained classification models.

ΔToF Data	Model	Validation Accuracy [%]	Testing Accuracy [%]
CC	Ensemble Subspace KNN	94.3	93.7
CWT	Ensemble Boosted Trees	97.2	98.0
STFT	Weighted KNN	85.6	85.7

https://doi.org/10.3390/instruments10010006

The performance of the models varies depending on the ΔToF estimation method. For CC data, the best performing model is an Ensemble Subspace KNN, which utilizes K-Nearest Neighbors (KNNs) enhanced by subspace and ensemble learning. It trains 30 KNN classifiers on different subsets of data with fewer features to improve performance. Considering CWT data, an Ensemble Boosted Trees model is used. This is composed of 30 decision trees trained using the AdaBoost algorithm. Finally, for STFT ΔToFs, the selected model is a Weighted KNN trained with 10 nearest neighbors. The weights for each of them are the squared inverse of the Euclidean distance.

Comparing the different ΔToF estimation methods, the model trained on the CWT data provides the best results with a classification accuracy of 98%. This is a consequence of the reduced variability of the ΔToFs, as shown in the previous section. For all trained models, the 5-fold cross-validation scheme avoids overfitting since the validation and testing accuracies are similar.

Figure 16 represents the confusion matrixes of the trained models in the testing phase, enabling us to better understand the classification capabilities of each trained model.

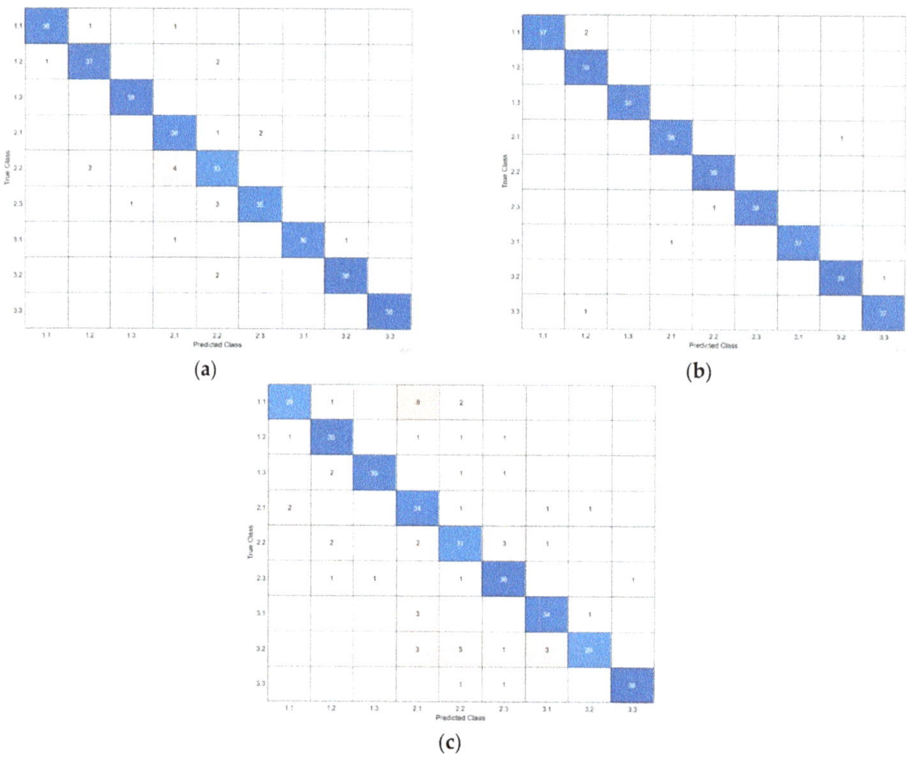

Figure 16. The confusion matrix in the testing phase for the classification models trained using CC (**a**), CWT (**b**), and STFT (**c**) ΔToF estimation algorithms. The blue squares on the diagonal and the corresponding numbers represent the correctly classified impacts for each zone. On the contrary, the red ones outside the diagonal are the number of impacts classified incorrectly for each zone.

From the confusion matrixes reported in Figure 16, for CC- and CWT-trained models, wrongly classified impacts are mostly localized in zones adjacent to the real one. The zone that is affected the most by this effect is 2.2, since it borders all the others. Considering the model trained using ΔToFs estimated with STFT (Figure 16c), there are several classification

https://doi.org/10.3390/instruments10010006

errors and many of them are not localized in nearby zones due to STFT frequency and time resolution.

3.2.1. Effect of Impact Point Layout

The effect of impact point layout is evaluated by removing the points on the borders of each zone. With this arrangement, errors due to the misclassification of impacts in zones close to each other are reduced. Table 3 summarizes the accuracy of the trained models using all impact points and the dataset without data on the borders.

As reported in Table 3, by removing the points on the borders, an increment in validation and testing accuracy is observed for all classification models. This is due to the reduction in misclassification errors in contiguous zones.

In Figure 17, the confusion matrixes of the models are reported.

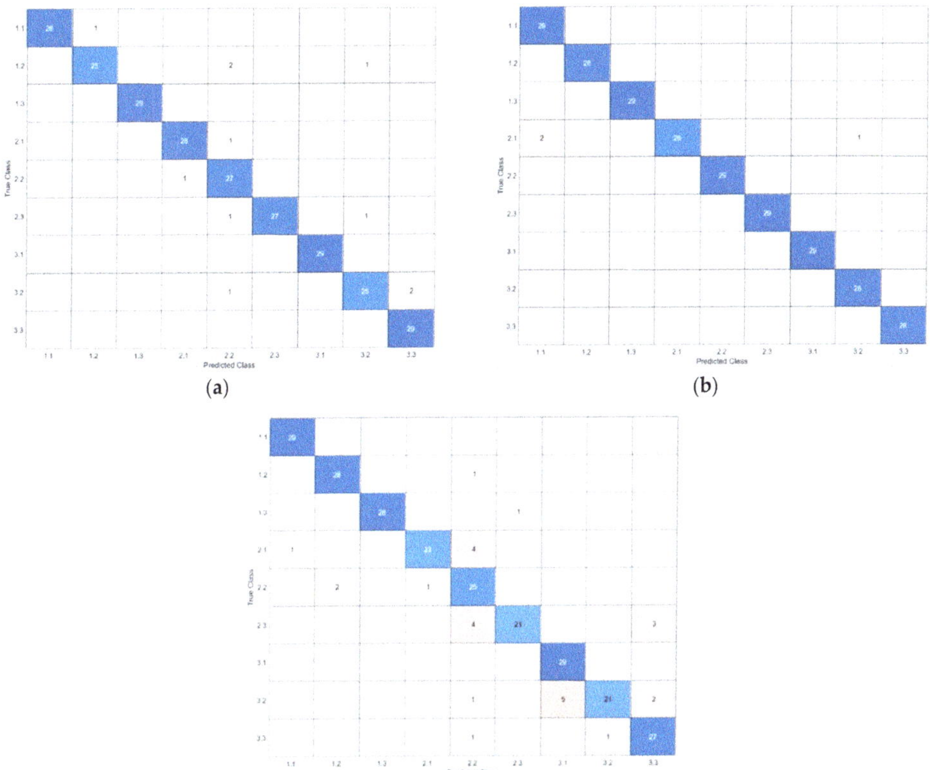

(a) (b) (c)

Figure 17. The confusion matrix of the classification models trained using CC (**a**), CWT (**b**), and STFT (**c**) ΔToF estimation algorithms after removing the impacts on the borders. The blue squares along the diagonal and the associated numbers represent the correctly classified impacts for each zone. In contrast, the red squares outside the diagonal indicate the number of impacts that were incorrectly classified for each zone.

As shown in Figure 17, there is a reduction in errors in contiguous zones for all models. However, as shown in Figure 17a, zone 2.2 still has different impacts classified in zones next to it. As shown in Figure 17c, eliminating the points on the border also improves the performance of the STFT model.

https://doi.org/10.3390/instruments10010006

Table 3. A summary of the accuracy of the classification models trained using all impact points and removing the points on borders.

	All Points		No Points on Borders	
ΔToF Data	Validation Accuracy [%]	Testing Accuracy [%]	Validation Accuracy [%]	Testing Accuracy [%]
CC	94.3	93.7	97.0	95.7
CWT	97.2	98.0	97.6	98.8
STFT	85.6	85.7	90.8	89.5

3.2.2. Effect of Number of Impact Points

The effect of the number of points used for training is evaluated by considering half impacts and a minimum distance between points of 35.35 mm.

Table 4 reports the comparison between the accuracies of the classification models trained using all impact points and using only half of the dataset.

Table 4. A summary of the accuracy of the classification models trained using all impact points and only half of the dataset.

	ALL POINTS		HALF POINTS	
ΔToF Data	Validation Accuracy [%]	Testing Accuracy [%]	Validation Accuracy [%]	Testing Accuracy [%]
CC	94.3	93.7	93.0	94.3
CWT	97.2	98.0	95.9	96.6
STFT	85.6	85.7	83.9	88.1

As shown in Table 4, no significative effects are observed due to the reduction in impact points.

3.2.3. Effect of Sampling Frequency

In Table 5, a summary of the accuracies of the models trained using subsampled signals is reported.

Table 5. A summary of the accuracy of the classification models trained using a dataset with different sampling frequencies.

	5 MHz		1 MHz		0.5 MHz	
ΔToF Data	Validation Accuracy [%]	Testing Accuracy [%]	Validation Accuracy [%]	Testing Accuracy [%]	Validation Accuracy [%]	Testing Accuracy [%]
CC	94.3	93.7	94.1	93.4	94.0	93.1
CWT	97.2	98.0	96.9	95.7	97.7	97.7
STFT	85.6	85.7	86.2	85.7	86.8	82.5

For all of the trained classification models, no significant variations are observed in the accuracy of the classification models. This is aspect is of great interest in terms of reducing the computational cost of the analysis, which required further analysis compared to previous works [10]. A previous study focused on a regression analysis for impact localization, which exploited the maximum sampling frequency of 5.0 MHz provided by the oscilloscope. Identification of the impacted area with lower acquisition and computational costs was confirmed. This represents a promising result for the optimization of the overall

https://doi.org/10.3390/instruments10010006

procedure, since the accuracy of classification is a global indicator, accounting for sampling frequency effects as well as the other effects analyzed.

3.2.4. Reproducibility Assessment

To verify the reproducibility of the analysis and further test the classification models, the above trained classification models are used with a different ΔToF dataset, obtained by means of tests carried out in a different laboratory under similar but not identical conditions; the models were trained with three datasets: one considering the total number of points, one excluding the points near the borders, and one using half of the points.

The results in terms of the accuracy of this analysis are summarized in Table 6.

Table 6. Classification accuracies using trained models on different datasets.

Model	TRAINING DATASET		
	All Points	**No Borders**	**Half Points**
CC	94.6	91.0	79.6
CWT	94.6	95.2	86.2
STFT	79.6	77.2	80.2

The models trained with the ΔToFs estimated with CWT proved to be the most accurate with respect to changes in the experimental setup. They showed satisfactory reproducibility even when challenged by a reduction in the number of training points. In the case of CC, a high sensitivity to the reduction in the number of points was highlighted. On the other hand, the STFT model showed less reproducibility when different experimental setup datasets were used as input. This outcome is probably due to the worse temporal resolution of the method.

Our results show that the models trained using CWT as the ΔToF estimator are more robust to reductions in training data and can classify impact signals acquired with a different setup while providing satisfactory accuracy. This promising outcome regarding the reproducibility of data processing and analysis is a relevant result for in-field implementation.

Therefore, as a final consideration, the aspects with the greatest impact on classification accuracy are the algorithm used to estimate the ToFs, the number of points used to train the ML models, the classification model chosen, the geometry of impact point distribution, and the balance of points across the classification areas; instead, a negligible effect was observed for acquisition frequency.

4. Conclusions

In this paper, the effects of different parameters were studied for classifying the localization of an impact on an aerospace component. The analysis of variability in different aspects of the methodology provides suggestions for improving the reproducibility of results. The main aspects that influence variability and bias errors in the results have been studied. These are related to the experimental procedure, such as the layout of the impact grid and its density, the shape of the impact areas to be classified, and the frequency of acquisition of data, and to the data processing technique, such as algorithms for estimating the differences in Times of Flight (ΔToFs) as well as ML classifiers and their parameters. Tests were carried out on an aluminum plate with four PZT sensors to acquire impact signals. Among CC, CWT, and STFT, it was found that CWT estimations of ΔToFs have the lowest mean variability, which is equal to 0.008 ms, and can be used as training features for different classification models. The optimized classification method shows a testing classification accuracy of 98%. Finally, these trained and improved models have

https://doi.org/10.3390/instruments10010006

been applied to datasets obtained in a different laboratory with the aim of studying the reproducibility of the results as well as the potential for an in-field transfer of the developed techniques. Our results show that the trained models exhibited an accuracy in the order of 95%, even though many testing conditions were changed such as the data acquisition system and the algorithm used to estimate ΔToFs, the data acquisition frequency, the operators, the impact spheres, and the laboratory environmental conditions. These results demonstrate the potential of the procedure not only for use with homogeneous materials but also for cases where more complex materials are such as in aerospace applications. Having quantitively assessed a possible benchmark for a data-driven ML approach for SHM application, the next step is to integrate it with advanced analytical models. This step is necessary for analyzing more complex components, such as composite plates made of carbon fiber or wing-shaped parts and reinforced components with screw holes. Indeed, the application of this procedure to more complex materials will be a further step in this project's work.

Author Contributions: Conceptualization, G.D., E.N., A.G., and F.N.; methodology, L.C., E.N., and A.G.; software, L.C. and F.N.; validation, L.C., A.G., and E.N.; formal analysis, L.C., E.N., A.G., and F.N.; investigation, L.C., E.N., A.G., and F.N.; resources, G.D. and G.S.; data curation, L.C. and F.N.; writing—original draft preparation, L.C. and E.N.; writing—review and editing, L.C., E.N., A.G., and F.N.; visualization, L.C.; supervision, G.D.; funding acquisition, G.D. and G.S. All authors have read and agreed to the published version of the manuscript.

Funding: This research received no external funding.

Data Availability Statement: The raw data supporting the conclusions of this article will be made available by the authors on request.

Conflicts of Interest: The authors declare no conflicts of interest.

Abbreviations

The following abbreviations are used in this manuscript:

SHM	Structural Health Monitoring
ML	Machine Learning
ToF	Time of Flight
NDT	Non-Destructive Testing
AE	Acoustic Emission
CC	Cross Correlation
CWT	Continuous Wavelet Transform
STFT	Short-Time Fourier Transform
SAE	Stacked Autoencoder
PWAS	Piezo Wafer Active sensor

Appendix A

ΔToF Estimation Method Pseudocodes

In this section, the pseudocodes for the ΔToF estimation methods are reported. All the algorithms consider only a pair of sensors.

https://doi.org/10.3390/instruments10010006

Algorithm 1: CrossCorrelationToF

1.

Input:
impactData

Output:
signalDelay: maxLagms

Algorithm:

```
BEGIN
// LOAD DATA
LOAD impactData

// TIME AXIS SETUP
SET sf TO 5 × 10^6// Sampling frequency in Hz
SET t TO (0 TO size(impactData, 1) - 1) × (1000 / sf)  // Time axis in ms

// DETECT CHANGE POINTS
FOR i FROM 1 TO 4 DO
SET changePt[i] TO findchangepts(impactData[:, i], "Statistic", "std")

SET minChangePt TO min(changePt)

// CONSIDER 0.6 ms OF SIGNAL
sensorData = impactData[minChangePt TO minChangePt + (0.6 / 1000 × sf), :]

// CREATE SIGNALS
SET signal1 TO sensorData[:, 1] // SENSOR 1 AND SENSOR 2 ARE CONSIDERED
SET signal2 TO sensorData[:, 2]

// DESIGN BANDPASS FILTER
SET lowCutoff TO 35,000
SET highCutoff TO 45,000
SET order TO 4
[b, a] = butter(order, [lowCutoff, highCutoff] / (sf / 2), 'bandpass')

// FILTER SIGNALS
signal1 = filter(b, a, signal1)
signal2 = filter(b, a, signal2)
// CALCULATE NORMALIZED CROSS-CORRELATION
[cross, lags] = xcorr(signal1, signal2, [], "normalized")
lagsms = lags * 1000 / sf //LAGS IN ms
// FIND MAXIMUM CROSS-CORRELATION AND LAG
[maxCross, maxIdx] = max(abs(cross))
SET maxLagms TO lagsms[maxIdx]

END
```

Figure A1. Pseudocode for ΔToF estimation using CC method.

https://doi.org/10.3390/instruments10010006

Algorithm 2: ContinuousWaveletToF

1.

Input:
ImpactData

Output:
signalDelay: deltaTof

Algorithm:

```
BEGIN
// LOAD DATA
LOAD impactData

// TIME AXIS SETUP
SET sf TO 5 × 10^6 // Sampling frequency in Hz
SET t TO (0 TO size(impactData, 1) - 1) × (1000 / sf) // Time axis in ms

// CREATE SIGNALS
SET signal1 TO impactData[:, 1] // SENSOR 1 AND SENSOR 2 ARE CON-SIDERED
SET signal2 TO impactData[:, 2]
// CALCULATE CONTINUOUS WAVELET TRANSFORM (CWT)
[cfs1, frq1] = cwt(signal1, "morse", fs, 'VoicesPerOctave', 16, 'FrequencyLimits', [20,000, 60,000])
[cfs2, frq2] = cwt(signal2, "morse", fs, 'VoicesPerOctave', 16, 'FrequencyLimits', [20,000, 60,000])

// FIND PEAKS IN CWT AT 40 kHz
[peaks1, peaks1Idx] = findpeaks(abs(cfs1(10, :)^2), "MinPeak-Height", 0.002)
[peaks2, peaks2Idx] = findpeaks(abs(cfs2(10, :)^2), "MinPeakHeight", 0.002)

// EXTRACT TIME OF FIRST PEAK
SET firstSignal1 TO t[peaks1Idx(1)]
SET firstSignal2 TO t[peaks2Idx(1)]

// CALCULATE DELTA TIME OF FLIGHT
SET deltaTof TO (firstSignal1 - firstSignal2)
END
```

Figure A2. Pseudocode for ΔToF estimation using CWT method.

https://doi.org/10.3390/instruments10010006

Algorithm 3: ShortTimeFourierTransformToF

1.

Input:
impactData

Output:
signalDelay: deltaTof

Algorithm:

```
BEGIN

// LOAD DATA
LOAD impactData

// TIME AXIS SETUP
SET sf TO 5 × 10^6 // Sampling frequency in Hz
SET t TO (0 TO size(impactData, 1) - 1) × (1000 / sf) // Time axis in ms

// SET PARAMETERS FOR STFT
SET M TO 25 // Window size of 0.1 ms
SET g TO hamming(m) //Humming window of size M
SET L TO 22 // Number of points for overlap
SET freq TO 40,000 // Frequency to extract in Hz

// CREATE SIGNALS
SET signal1 TO impactData[:, 1] // SENSOR 1 AND SENSOR 2 ARE CON-SIDERED
SET signal2 TO impactData[:, 2]

// CALCULATE STFT
[s1, f1, t1, ps1] = spectrogram(signal1, g, 0, M, fs)
[s2, f2, t2, ps2] = spectrogram(signal2, g, 0, M, fs)

// FIND PEAKS IN STFT
[peaks1, peaks1Idx] = findpeaks(abs(s1(f1 == freq, :)^2), "MinPeakHeight", 0.1)
[peaks2, peaks2Idx] = findpeaks(abs(s2(f1 == freq, :)^2), "MinPeakHeight", 0.1)

// EXTRACT TIMES OF FIRST PEAK
SET firstSignal1 TO t1(peaks1Idx(1)) × 1000 // Convert to ms
SET firstSignal2 TO t2(peaks2Idx(1)) × 1000 // Convert to ms

// CALCULATE TIME DIFFERENCE
SET deltaTof TO (firstSignal1 - firstSignal2)

END
```

Figure A3. Pseudocode for ΔToF estimation using STFT method.

References

1. Ricci, F.; Mal, A.K.; Monaco, E.; Maio, L.; Boffa, N.D.; Di Palma, M.; Lecce, L. Guided Waves in Layered Plate with Delaminations. In *Proceedings of the EWSHM—7th European Workshop on Structural Health Monitoring*; Le Cam, V., Mevel, L., Schoefs, F., Eds.; IFFSTTAR, Inria, Université de Nantes: Nantes, France, 2014.
2. Monaco, E.; Boffa, N.D.; Memmolo, V.; Ricci, F.; Maio, L. Detecting Delaminations and Disbondings on Full-Scale Wing Composite Panel by Guided Waves Based SHM System. In *Health Monitoring of Structural and Biological Systems 2016*; SPIE: Bellingham, WA, USA, 2016; Volume 9805, pp. 255–262.
3. Boccardi, S.; Boffa, N.D.; Carlomagno, G.M.; Maio, L.; Meola, C.; Ricci, F. Infrared Thermography and Ultrasonics to Evaluate Composite Materials for Aeronautical Applications. *J. Phys. Conf. Ser.* **2015**, *658*, 012007. [CrossRef]
4. Boccardi, S.; Boffa, N.D.; Carlomagno, G.M.; Del Core, G.; Meola, C.; Russo, P.; Simeoli, G. Inline Monitoring of Basalt-Based Composites under Impact Tests. *Compos. Struct.* **2019**, *210*, 152–158. [CrossRef]
5. Natale, E.; Gaspari, A.; Chiominto, L.; D'Emilia, G.; Stamopoulos, A.G. Morphological Analysis of As-Manufactured Filament Wound Composite Cylinders Using Contact and Non-Contact Inspection Methods. *Eng. Fail. Anal.* **2024**, *158*, 108011. [CrossRef]
6. D'Emilia, G.; Gaspari, A.; Natale, E.; Stamopoulos, A.G.; Di Ilio, A. Experimental and Numerical Analysis of the Defects Induced by the Thermoforming Process on Woven Textile Thermoplastic Composites. *Eng. Fail. Anal.* **2022**, *135*, 106093. [CrossRef]

https://doi.org/10.3390/instruments10010006

7. Beard, S.J.; Kumar, A.; Qing, X.; Chan, H.L.; Zhang, C.; Ooi, T.K. Practical Issues in Real-World Implementation of Structural Health Monitoring Systems. In *Smart Structures and Materials 2005: Industrial and Commercial Applications of Smart Structures Technologies*; SPIE: Bellingham, WA, USA, 2005; Volume 5762, pp. 196–203.
8. Malekloo, A.; Ozer, E.; AlHamaydeh, M.; Girolami, M. Machine Learning and Structural Health Monitoring Overview with Emerging Technology and High-Dimensional Data Source Highlights. *Struct. Health Monit.* **2022**, *21*, 1906–1955. [CrossRef]
9. Ciang, C.C.; Lee, J.-R.; Bang, H.-J. Structural Health Monitoring for a Wind Turbine System: A Review of Damage Detection Methods. *Meas. Sci. Technol.* **2008**, *19*, 122001. [CrossRef]
10. Dipietrangelo, F.; Nicassio, F.; Scarselli, G. Structural Health Monitoring for Impact Localisation via Machine Learning. *Mech. Syst. Signal Process.* **2023**, *183*, 109621. [CrossRef]
11. Kosova, F.; Altay, Ö.; Ünver, H.Ö. Structural Health Monitoring in Aviation: A Comprehensive Review and Future Directions for Machine Learning. *Nondestruct. Test. Eval.* **2025**, *40*, 1–60. [CrossRef]
12. Cha, J.-H.; Kim, Y.; Sathish Kumar, S.K.; Choi, C.; Kim, C.-G. Ultra-High-Molecular-Weight Polyethylene as a Hypervelocity Impact Shielding Material for Space Structures. *Acta Astronaut.* **2020**, *168*, 182–190. [CrossRef]
13. Jaussaud, G.; Rebufa, J.; Fournier, M.; Logeais, M.; Bencheikh, N.; Rébillat, M.; Guskov, M. Improving Lamb Wave Detection for SHM Using a Dedicated LWDS Electronics. In *11th International Symposium on NDT in Aerospace*; COFREND: Paris, France, 2019; Volume 25, pp. 1–7.
14. Ciampa, F.; Meo, M. A New Algorithm for Acoustic Emission Localization and Flexural Group Velocity Determination in Anisotropic Structures. *Compos. Part A Appl. Sci. Manuf.* **2010**, *41*, 1777–1786. [CrossRef]
15. Kundu, T. Acoustic Source Localization. *Ultrasonics* **2014**, *54*, 25–38. [CrossRef] [PubMed]
16. Marino-Merlo, E.; Bulletti, A.; Giannelli, P.; Calzolai, M.; Capineri, L.; Marino-Merlo, E.; Bulletti, A.; Giannelli, P.; Calzolai, M.; Capineri, L. Analysis of Errors in the Estimation of Impact Positions in Plate-Like Structure through the Triangulation Formula by Piezoelectric Sensors Monitoring. *Sensors* **2018**, *18*, 3426. [CrossRef] [PubMed]
17. Ebrahimkhanlou, A.; Dubuc, B.; Salamone, S. A Generalizable Deep Learning Framework for Localizing and Characterizing Acoustic Emission Sources in Riveted Metallic Panels. *Mech. Syst. Signal Process.* **2019**, *130*, 248–272. [CrossRef]
18. Cuomo, S.; De Simone, M.E.; Andreades, C.; Ciampa, F.; Meo, M. Machine Learning for Impact Detection on Composite Structures. *Mater. Today Proc.* **2021**, *34*, 93–98. [CrossRef]
19. Ai, L.; Soltangharaei, V.; Bayat, M.; Van Tooren, M.; Ziehl, P. Detection of Impact on Aircraft Composite Structure Using Machine Learning Techniques. *Meas. Sci. Technol.* **2021**, *32*, 084013. [CrossRef]
20. Chiominto, L.; Natale, E.; D'Emilia, G.; Grieco, S.A.; Prato, A.; Facello, A.; Schiavi, A. Responsiveness and Precision of Digital IMUs under Linear and Curvilinear Motion Conditions for Local Navigation and Positioning in Advanced Smart Mobility. *Micromachines* **2024**, *15*, 727. [CrossRef] [PubMed]
21. Chiominto, L.; D'Emilia, G.; Dipietrangelo, F.; Gaspari, A.; Natale, E.; Nicassio, F.; Scarselli, G. Integrated Sensors for Structural Health Monitoring: Groundwork for an Interlaboratory Comparison. In *2025 IEEE International Workshop on Metrology for Industry 4.0 & IoT (MetroInd4.0 & IoT)*; IEEE: New York, NY, USA, 2025; pp. 45–50.
22. Piezoceramics Disks. Available online: https://www.physikinstrumente.com/en/products/piezoelectric-transducers-actuators/disks-rods-and-cylinders (accessed on 31 October 2025).
23. findchangepts. MATLAB Documentation. Available online: https://it.mathworks.com/help/signal/ref/findchangepts.html (accessed on 31 October 2025).
24. xcorr. MATLAB Documentation. Available online: https://it.mathworks.com/help/matlab/ref/xcorr.html (accessed on 31 October 2025).
25. Martinez-Ríos, E.A.; Bustamante-Bello, R.; Navarro-Tuch, S.; Perez-Meana, H. Applications of the Generalized Morse Wavelets: A Review. *IEEE Access* **2023**, *11*, 667–688. [CrossRef]
26. findpeaks. MATLAB Documentation. Available online: https://it.mathworks.com/help/signal/ref/findpeaks.html?s_tid=srchtitle_support_results_1_findpeaks (accessed on 31 October 2025).
27. *ISO/IEC, Guide 98-3:2008*; Uncertainty of Measurement—Part 3: Guide to the Expression of Uncertainty in Measurement. ISO: Geneva, Switzerland, 2018.
28. resample. MATLAB Documentation. Available online: https://it.mathworks.com/help/signal/ref/resample.html?s_tid=srchtitle_support_results_1_resample (accessed on 31 October 2025).

https://doi.org/10.3390/instruments10010006

instruments

MDPI

Article

Evaluation of Gap and Flush Inspection Algorithms in a Portable Laser Line Triangulation System Through Measurement System Analysis (MSA)

Guerino Gianfranco Paolini [1], Sara Casaccia [1,*], Matteo Nisi [2], Cristina Cristalli [2] and Nicola Paone [1]

[1] Dipartimento di Ingegneria Industriale e Scienze Matematiche, Università Politecnica delle Marche, 60131 Ancona, Italy; g.g.paolini@pm.univpm.it (G.G.P.); n.paone@staff.univpm.it (N.P.)
[2] U-Sense IT Srl, 60035 Jesi, Italy; nisi@u-sense.it (M.N.); cristalli@u-sense.it (C.C.)
* Correspondence: s.casaccia@staff.univpm.it

Abstract

The shift toward Industry 5.0 places human-centred and digitally integrated metrology at the core of modern manufacturing, particularly in the automotive sector, where portable Laser Line Triangulation (LLT) systems must combine accuracy with operator usability. This study addresses the challenge of operator-induced variability by evaluating how algorithmic strategies and mechanical support features jointly influence the performance of a portable LLT device derived from the G3F sensor. A comprehensive Measurement System Analysis was performed to compare three feature extraction algorithms—GC, FIR, and Steger—and to assess the effect of a masking device designed to improve mechanical alignment during manual measurements. The results highlight distinct algorithm-dependent behaviours in terms of repeatability, reproducibility, and computational efficiency. More sophisticated algorithms demonstrate improved sensitivity and feature localisation under controlled conditions, whereas simpler gradient-based strategies provide more stable performance and shorter processing times when measurement conditions deviate from the ideal. These differences indicate a trade-off between algorithmic complexity and operational robustness that is particularly relevant for portable, operator-assisted metrology. The presence of mechanical alignment aids was found to contribute to improved measurement consistency across all algorithms. Overall, the findings highlight the need for an integrated co-design of algorithms, calibration procedures, and ergonomic aids to enhance repeatability and support operator-friendly LLT systems aligned with Industry 5.0 principles.

Keywords: laser line triangulation system; measurement system analysis; gap and flush measurement

1. Introduction

The evolution of manufacturing towards intelligent, adaptive, and human-centric paradigms has placed measurement technologies at the heart of industrial transformation. Industry 5.0, emphasizing the synergy between human expertise and advanced automation, requires quality control systems that are not only precise but also flexible, resilient, and capable of supporting customization and safety across diverse production environments [1,2].

This evolution is closely intertwined with the broader transition toward digitalization [3], which is redefining industrial processes across sectors through smart technologies, interconnected systems, and data-driven decision-making. Measurement systems are no longer isolated or purely manual instruments; they are becoming integrated platforms capable of real-time analysis, remote monitoring, and seamless communication with other digital assets. By combining advanced measurement capabilities with intelligent digital frameworks, manufacturers can achieve greater efficiency, traceability, and adaptability—laying the foundation for production environments that fully embody the principles of Industry 5.0 [4].

Within this broader transformation, the automotive sector exemplifies how digital metrology is reshaping traditional quality control practices. The integration of advanced sensors and automated inspection routines has become essential to meet the increasingly stringent aesthetic and functional requirements of modern vehicles [5]. Dimensional checks such as gap and flush—used to assess the alignment and surface continuity between adjacent body components—play a vital role in ensuring both visual appeal and aerodynamic performance. Their accurate detection is now a critical step in digitalized assembly lines, where precision and repeatability must coexist with speed and flexibility [6]. Gap refers to the lateral distance between two neighbouring surfaces, while flush quantifies the vertical offset between them. Historically, these measurements were performed manually using gauges and visual inspection, but such methods are increasingly inadequate in the face of tighter tolerances and the need for traceability. Laser triangulation has emerged as a robust alternative, offering non-contact, high-resolution data acquisition suitable for both in-line and off-line applications [7].

Recent research has explored various implementations of portable triangulation systems [8], including smartphone-integrated sensors [9–11] and robotic-mounted devices [12]. These solutions have demonstrated promising results in terms of mobility and integration with digital ecosystems [13]. However, a persistent challenge remains: the variability introduced by human operators when using handheld systems. While systematic errors can be minimized through calibration and design, operator-induced deviations continue to affect measurement reliability, especially in manual modes [14].

In manual measurement contexts, the role of the operator becomes a critical variable influencing the reliability of the acquired data. Even when using high-precision instruments, the absence of standardized handling procedures or real-time feedback mechanisms can lead to inconsistencies in positioning, orientation, and the interpretation of results. These uncertainties, often subtle and difficult to detect, can compromise the validity of the analysis and hinder process optimization [15].

To mitigate these risks, it is essential to equip operators with advanced support techniques that enhance measurement repeatability and reduce subjective influence. This includes the integration of algorithmic guidance systems, ergonomic fixtures, and intelligent interfaces capable of assisting the user during acquisition. By embedding such tools into the measurement workflow, the system can actively compensate for human variability, ensuring that the data collected reflect the true geometry of the components rather than the nuances of operator behaviour.

In response to this challenge, the present study builds on the existing G3F sensor, a portable Laser Line Triangulation (LLT) system previously reported in the literature for automotive applications [16], and introduces a novel contribution: a comparative analysis of three dedicated algorithms for gap and flush measurement. The key innovation of this work lies in evaluating these algorithms through a rigorous Measurement System Analysis (MSA), which systematically assesses their accuracy, repeatability, and reproducibility under varying operational conditions. While the G3F system itself is already established, this study provides new insights into the relative performance of the algorithms, highlighting how differences in mathematical modelling, signal processing strategies, and sensitivity to

https://doi.org/10.3390/instruments10010007

environmental factors affect measurement reliability. This MSA-based comparison offers a robust framework for benchmarking laser-based gap and flush measurement techniques in real-world automotive scenarios.

Beyond this algorithmic comparison, this study also examines a modification of the G3F system in which a masking device is mounted to facilitate measurement by the operator. By contrasting the standard G3F sensor with the masked version, the research assesses not only algorithmic performance but also how hardware adaptations can influence measurement ease, precision, and reliability. This dual-level evaluation—algorithmic and hardware-assisted—provides a comprehensive framework for benchmarking laser-based gap and flush measurement techniques in real-world automotive scenarios.

Ultimately, the results of these analyses will provide a clear picture of both the measurement system's performance and the operator's influence on the process. By quantifying these effects and comparing algorithmic strategies, this study aims to advance the integration of intelligent measurement systems into human-centred manufacturing workflows, contributing to more resilient, adaptive, and user-friendly metrology solutions fully aligned with the principles of Industry 5.0.

This paper is organized as follows: Section 2 describes a summary of related works to select which algorithm to choose for deeper study; Section 3 describes the materials and methods; Section 4 presents the results; Section 5 presents the limitations of this study and conclusion.

2. Related Works for Algorithm Selection

The main novelty of this study is the systematic evaluation of the trade-off between measurement accuracy and processing time on an embedded platform using well-established stripe extraction algorithms. Although numerous works address laser stripe detection for Laser Line Triangulation (LLT) systems, quantitative studies that jointly assess metrological performance and computational cost within a unified benchmarking framework remain scarce, particularly for portable and embedded industrial applications. In the existing literature, algorithm selection is often driven either by accuracy considerations alone or by application-specific constraints, without an explicit comparison of how different methods balance precision, robustness, and execution time on the same hardware platform. As a result, practical guidelines for algorithm selection under real-time and resource-constrained conditions are still limited. This study aims to fill this gap by providing a fair and reproducible benchmark in which accuracy, repeatability, reproducibility, and processing time are evaluated simultaneously.

In particular, a review of the literature [17] reveals that stripe extraction algorithms for Laser Line Triangulation (LLT) system are generally categorized based on their approach to finding the centre, balancing accuracy against computational cost. The most common and fundamental methods cited are centroid-based algorithms (like the Grey Centroid—GC) [18] and Hessian-based algorithms (like the Steger method) [19]. Many recent studies focus on developing new algorithms and Artificial Intelligence (AI) techniques that modify or combine these methods to achieve higher accuracy. However, these enhancements often introduce further mathematical operations, increasing the processing time, which is unsuitable for the embedded system targeted in this work. An additional intermediate algorithm in terms of computational cost can join with the GC and Steger methods to perform the benchmark task, namely the Finite Impulse Response (FIR) algorithm [20]. Details of each algorithm chosen is described in Section 3.5 (Algorithms).

A recent study [14] employed the GC algorithm and performed a Gauge Repeatability and Reproducibility (GRR) analysis, reporting a gap GRR of 0.099 mm and a flush GRR of 0.075 mm on a real vehicle. In that work, the algorithms were executed on a smartphone

https://doi.org/10.3390/instruments10010007

platform, and the measured cycle time per point was reported as 11 s; however, no details are provided on how this time was obtained. The main differences between [14] and the present study concern the hardware platform (professional embedded system versus smartphone), the filtering kernel size, the thresholding method, and the feature extraction algorithm applied to the profile. Further details of the proposed methodology are reported in Section 3 (Materials and Methods).

In recent years, machine learning (ML) and AI have emerged as promising approaches to improve the accuracy and adaptability of LLT systems. Instead of relying purely on analytical models, AI-based methods can learn the mapping between image intensity distributions and stripe centre positions directly from data. Neural network architectures, such as convolutional neural networks (CNNs), have been trained to detect laser stripe centres with sub-pixel accuracy even under challenging conditions like varying surface reflectivity, ambient illumination, or partial occlusion [21]. Other approaches exploit deep regression models and autoencoders to denoise and enhance stripe profiles prior to extraction, thereby reducing sensitivity to sensor noise and optical distortions. Recent studies further show that integrating a neural network segmentation stage to isolate the stripe region, followed by sub-pixel centre extraction on the cleaned area, achieves higher accuracy and lower computational cost compared to full-image Hessian methods [22]. This combination of semantic segmentation and geometric refinement enables robust operation under difficult optical conditions while concentrating computation on the relevant region and minimizing stray light artifacts.

While these AI-driven solutions can significantly improve robustness and accuracy, they often require substantial computational resources and large training datasets, making them less suitable for real-time embedded implementations. Nevertheless, they represent a growing research direction for adaptive and self-calibrating triangulation sensors that can outperform traditional model-based algorithms in uncontrolled environments.

3. Materials and Methods

3.1. Methodology

The methodology adopted in this study has been structured to provide a clear and systematic assessment of the portable LLT system under investigation. The block diagram reported in Figure 1 shows the measurement procedure adopted. First, the sensing technology related to LLT is introduced, with a detailed description of the G3F sensor and the experimental test developed for gap and flush measurements on an automotive body. Subsequently, the image processing pipeline is presented, including the image pre-processing phase and the gap and flush extraction stage, where three different algorithms are introduced and discussed, highlighting their operating principles and implementation within the measurement workflow.

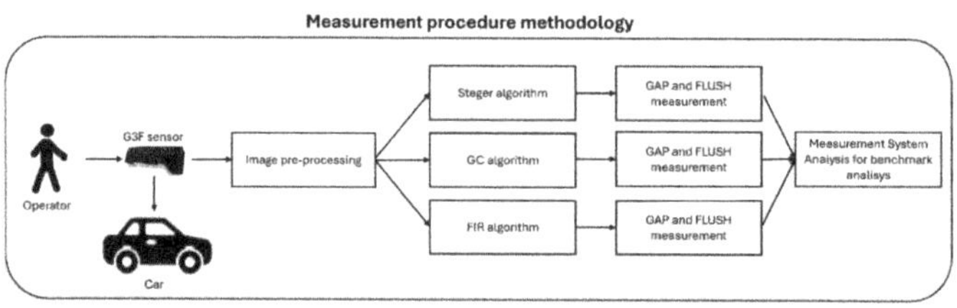

Figure 1. Block diagram describing measurement system methodology.

Finally, a Measurement System Analysis (MSA) is performed to assess the performance of the proposed algorithms, evaluating measurement uncertainty, repeatability,

https://doi.org/10.3390/instruments10010007

reproducibility, and processing time, in order to compare their accuracy and computational efficiency [23].

This structure enables a comprehensive understanding of how hardware setup, measurement system performance, and algorithmic strategies jointly influence the reliability of gap and flush measurements in automotive applications. Within this framework, the flush value is equal to the distance between the straight line of a reference surface and the edge point on the other side; in practice, both the reference line and the edge point are geometric features that must be detected. The gap value is evaluated as the distance between the circles of surfaces; these circles are likewise geometric features extracted from the measurement data. Figure 2 shows two adjoining components (solid lines) together with these geometric definitions of gap and flush [9].

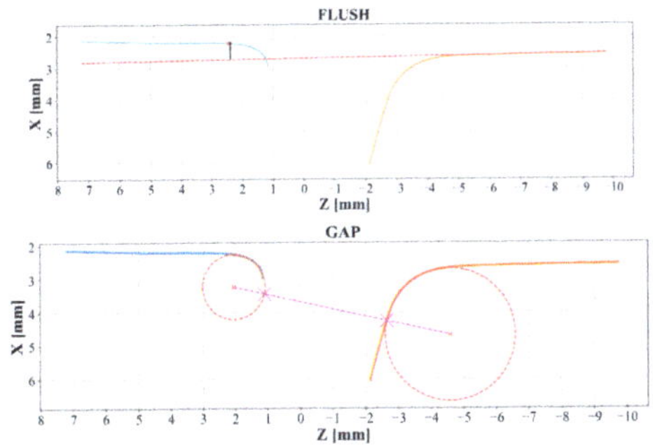

Figure 2. Flush and Gap definitions. The left profile is identified in blue and the right profile in orange. The extracted features are highlighted in red: the circle for the gap and the straight line for the flush. The distance between the two identified gap centers, shown in purple, is included to improve visualization.

3.2. Core Technology: Laser Stripe Detection and Centre Estimation

The core task of an LLT sensor is to accurately determine the centre of the laser stripe in each column of the image, as this information forms the basis for the subsequent reconstruction of the observed object's profile. The laser beam profile can typically be approximated by a Gaussian distribution; therefore, along each image column j, the intensity of the laser line in the image can be represented as $I(i,j)$, where i denotes the pixel position within the column. After applying an appropriate intensity threshold, only pixels whose values exceed this threshold are retained as part of the laser stripe, as demonstrated in Figure 3 for (1). These selected pixels define the region of interest on which sub-pixel localization algorithms are applied to compute the stripe centre.

$$I(i,j) : i \in [k;n] \ \& \ I(i,j) \geq threshold. \quad (1)$$

In this context, it is essential to determine which laser stripe extraction algorithm is most suitable for real-time processing on an embedded system. The analysis will focus on several well-known algorithms, evaluating their performance in terms of precision and processing time when executed on a low-power CPU. The objective is to quantify the trade-off between measurement uncertainty and computational cost for a general-purpose LLT sensor controlled by a System-on-Module (SoM) platform.

Additionally, this study aims to assess how ergonomic modifications of the sensor influence its overall performance, and how these effects compare to the performance variations introduced by different centre detection algorithms.

https://doi.org/10.3390/instruments10010007

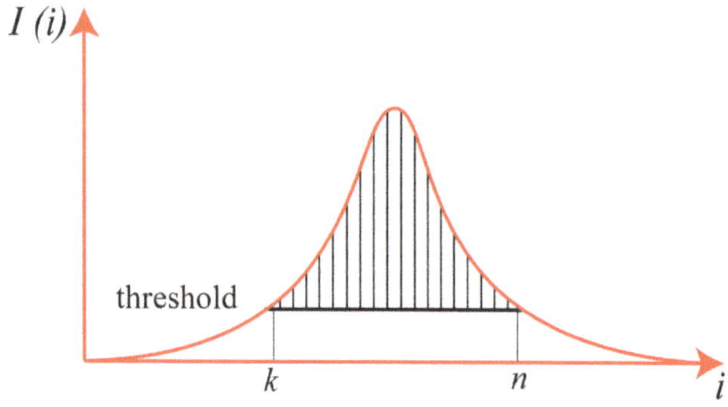

Figure 3. Gaussian profile with a threshold that determines where the laser stripe starts.

3.3. Measurement Setup

In this work, the G3F sensor from U-Sense.IT srl is used (Figure 4a), which is a manual, wireless, portable laser profilometer. To obtain the LLT Field of View (FOV), the components shown in Figure 5 are required:

- A laser source (blue in the figure) that projects a thin, straight-light plane;
- A camera, composed of a CMOS sensor combined with a lens system (represented as the yellow pyramid frustum).

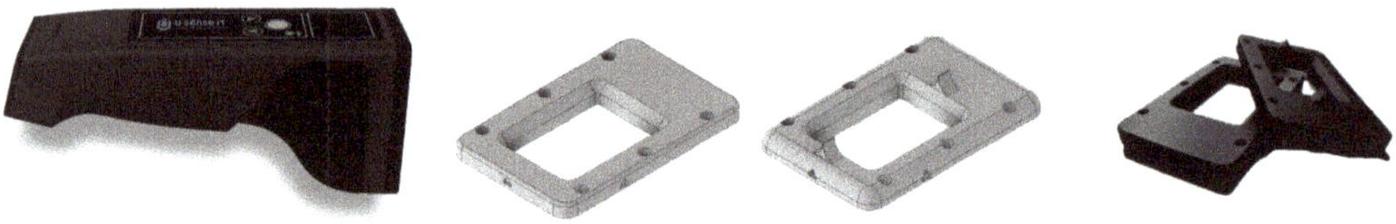

(**a**) (**b**) (**c**)

Figure 4. (**a**) G3F sensor; (**b**) standard soft sensor front panel (**left**) and prototype panel with alignment teeth (**right**); (**c**) real front panels.

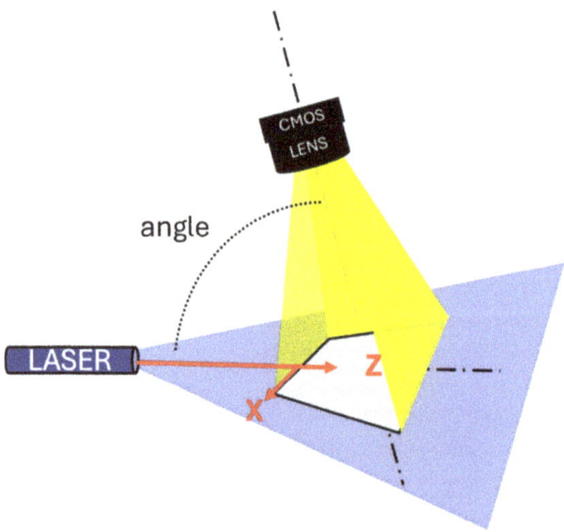

Figure 5. LLT FOV definition (grey) as interception of camera pyramid frustum FOV (yellow) and laser plane (bright blue).

https://doi.org/10.3390/instruments10010007

An angle is formed between the optical axis of the camera and that of the laser. The interception of the camera's FOV with the laser plane defines the sensor's trapezoidal measurement area [24].

The G3F sensor interacts with the contact surface through a soft front panel positioned in front of the optical assembly (the design of the soft panel is shown on the left in Figure 4b,c). The repeatability of the contact point has the greatest influence on the overall measurement uncertainty, as previously investigated in [16]. In this paper, an improvement in performance achieved through a more ergonomic and cost-effective design is presented. In this revised design, alignment teeth for both the gap and the flush have been added to the front panel, as shown on the right side in Figure 4b,c. This modification represents a low-cost design enhancement that improves positioning repeatability and user handling.

Table 1 describes the G3F FOV, defined along the Z-axis from the trapezoidal minor base at Z = 0 mm to the major base at Z = 12 mm. The plane at Z = 0 mm corresponds to measurements performed in contact with the sensor front panel, while the major base is located 12 mm away from it. Figure 6a shows the orientation of the axes with respect to the sensor front panel, and Figure 6b shows the sensor's FOV.

Table 1. Sensor's FOV description.

	Z = 0 mm	**Z = 12 mm**
X range	±8.5 mm	±11 mm
X Resolution	7 µm	9 µm
Z Resolution	9 µm	9 µm

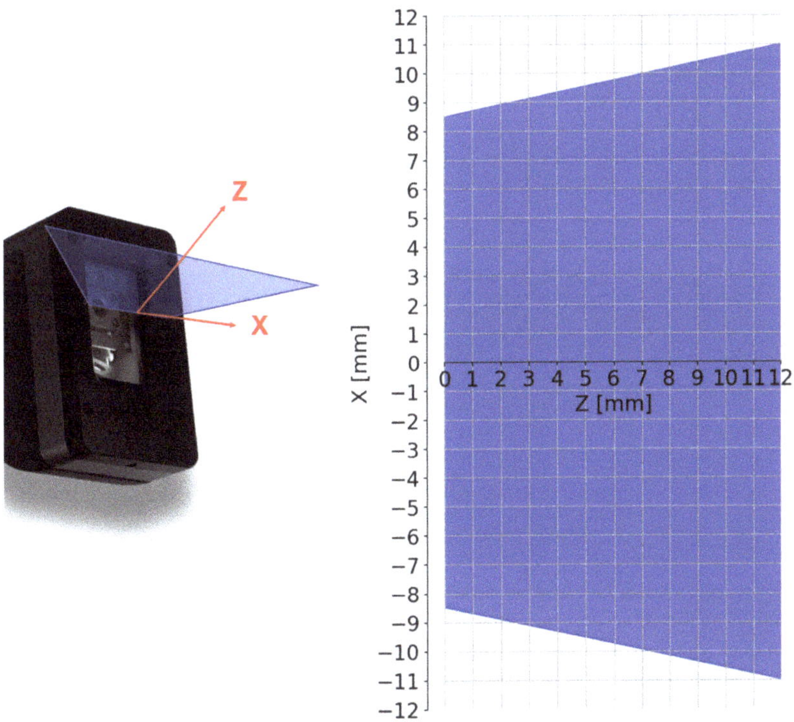

(**a**) (**b**)

Figure 6. (**a**) The orientation of the FOV on the sensor; (**b**) the shape of the sensor FOV in millimetres.

https://doi.org/10.3390/instruments10010007

3.4. Measurement Test Procedure

To perform this benchmark task in a controlled manner, a single set of images was used. Images were acquired from 14 different points on the same car, 7 for each side as shown in Figure 7, by four different operators. Each operator acquired the same point 3 times in a random order. This set was acquired under two distinct working conditions ("with" and "without alignment teeth"). Using the same image set ensures that all three algorithms are tested against the exact same input data, allowing for a direct comparison of their GRR performance and processing speed.

Figure 7. Seven measurement sequence points are shown in green on the car's left side. On the right side, the same seven points are positioned in corresponding locations.

As reported in the manual [23], a minimum number of 10 points should be acquired. In the same manual, the examples are recorded by two or three operators. To increase the statistical relevance, we conducted the study with four operators chosen considering four distinct levels of technology knowledge: producer knowledge, high technology knowledge, medium technology knowledge, and basic technology knowledge.

3.5. Algorithms

The measurement obtained by LLT is based on image acquisition. This context makes it possible to leverage standard Computer Vision pre-processing strategies to improve the robustness and accuracy of the data, primarily by reducing noise.

However, the industrial environment introduces numerous challenges that complicate the reliable extraction of the laser stripe centre. The primary challenges include the following:

- Image noise: requiring de-noising strategies (e.g., Gaussian, mean, or median filtering [17]);
- Surface irregularities: such as specular reflections or abrupt changes in colour or material;
- Laser stripe variations: where surface irregularities cause fluctuations in the stripe's thickness and intensity, producing heterogeneous regions in the image.

There are three main processing stages that influence the final measurement value. The first stage, image pre-processing, addresses noise and background identification [17]. The output of this stage is a denoised and properly cropped image, optimized for subsequent analysis. The second stage comprises laser stripe extraction. After applying the calibration functions, this stage produces one or more profiles representing the reconstructed geometry. To extract the laser stripe centreline with high precision, we employ the GC, Steger, and FIR algorithms. Finally, the third stage consists of feature extraction from the obtained profiles. In this phase, the relevant geometric features are computed to produce the final measurement output.

https://doi.org/10.3390/instruments10010007

3.5.1. Image Pre-Processing

After analysing the noise present in the images acquired by the sensor, it is observed that the noise is predominantly impulsive rather than Gaussian. This is evident in Figure 8, where high-intensity white dots appear in proximity to darker dots. This type of noise, known as *salt-and-pepper noise*, is characterized by frequent dark spots in bright areas and bright spots in dark areas across the image. Given these characteristics, the median filter emerges as the most suitable choice, as confirmed in the literature [25,26]; in this work, a median filter with a 5×5 kernel is adopted and the resulting improvements in the uniformity of the laser stripe are illustrated on the right in Figure 9.

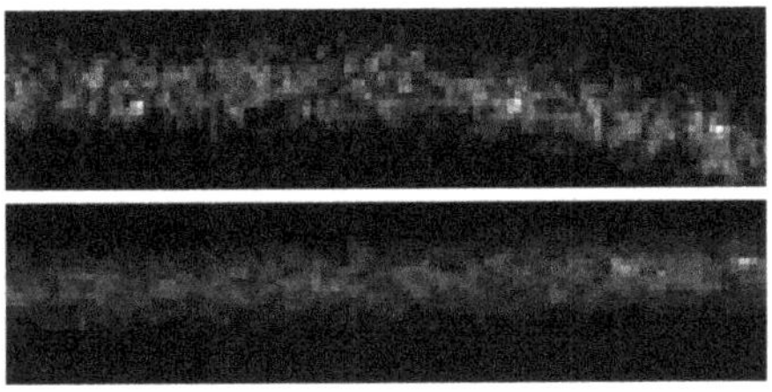

Figure 8. Two different images acquired by the sensor zoomed in on the laser stripe. It is possible to see bright points immersed in the laser stripe next to dark points, showing impulsive noise.

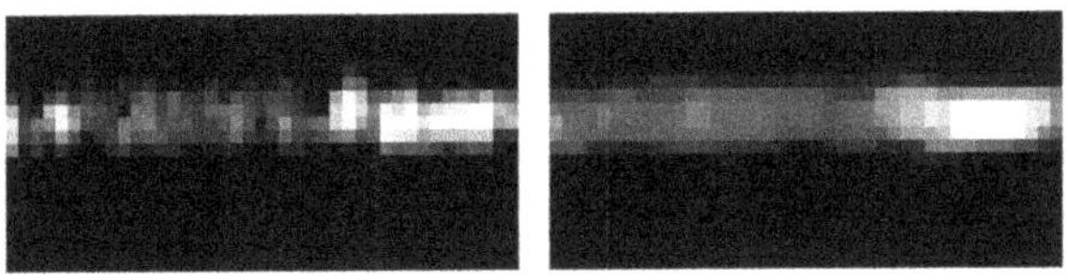

Figure 9. The left shows the raw portion of the image, and the right shows the same portion after using the median filter.

To increase the speed of laser stripe extraction and improve detection, the Otsu thresholding method is applied to set to zero all points outside the ROI [22]. The Otsu method is an automatic thresholding technique that assumes that the image is composed of two main pixel classes: foreground and background. For each possible threshold value, the image is divided into these two classes and the inter-class variance is computed, which measures how well separated the grey-level distributions of the two classes are. The optimal threshold maximizes this variance, as it corresponds to the best possible separation between foreground and background in terms of pixel intensity, and it is a standard algorithm available within openCV library and well explained in the documentation available in [27]. The positions of the first and last pixels with intensity $I(i,j) \neq 0$ (above the Otsu threshold) in each column are located by scanning the image from top to bottom, and the corresponding row indices are recorded as i_1 and i_2.

To enhance the stability of subsequent stripe centre extraction, the ROI is obtained considering only row indexes where pixel intensities are non-zero and is then expanded by 10 pixels both upward and downward. Therefore, the upper and lower boundaries of the ROI are defined as $(i_1 - 10)$ and $(i_2 + 10)$, respectively.

3.5.2. Stripe Extraction Algorithms: GC, Steger, and FIR

The GC algorithm finds the $i_{stripe-j}$ for each column j, i.e., the i row in the j column where the laser stripe is founded, following the equation for each consecutive pixel with intensity $I(i,j)$ that exceeds the threshold, as shown in Figure 10, and following Formula

(2) with sub-pixel accuracy that calculates the weighted average row above the threshold, where the weights are the levels of intensities $I(i,j)$:

$$i_{stripe-j} = \frac{\sum_{i=k}^{n} I(i,j)\, i}{\sum_{i=k}^{n} I(i,j)}. \tag{2}$$

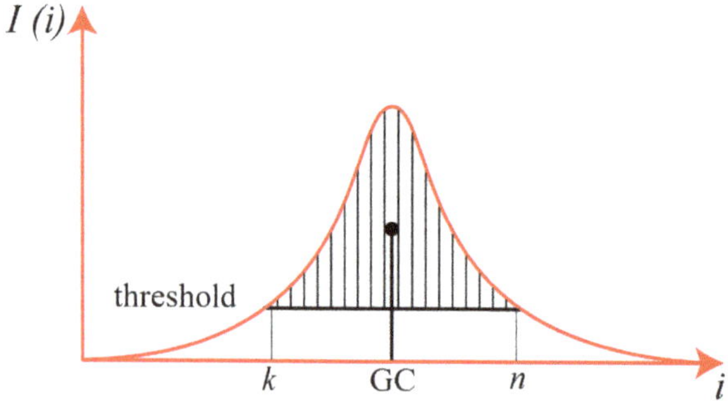

Figure 10. Representation of GC method for laser stripe extraction. Only values above threshold are considered. There are n −k black vertical lines representing pixels i rows above threshold.

The Steger method requires knowledge of the input parameter σ, which is related to stripe width and is used in the Gaussian filtering step. The algorithm then computes the Hessian matrix of the image and calculates its eigenvalues to identify the stripe profile with sub-pixel accuracy using the Taylor expansion along the normal direction, as described by Steger [19] and Zaho et. al. [28].

The Finite Impulse Response (FIR) algorithm is selected as an intermediate solution between GC and Steger, as it relies on a per-column convolution that is more computationally demanding than the Grey Centroid method but less complex than Steger. An important advantage of the FIR approach is that it applies a low-pass filter $h(i)$ during the computation of the derivative along each column j. It is possible to demonstrate that, considering vector $x(i)$, a filtered vector obtained with the convolution between the filter and the original vector (3) is equal to the vector convolved with the derived filter, as shown in (4), which means that the derivate must be calculated only once by the program after the filter definition [20]. Then, it is possible to calculate the i row with sub-pixel precision by finding the zero crossing, as shown in Figure 11, following Equation (5) where $p_0 = (x_0; y_0)$ is the last positive point of $dy(i)/di$ before the zero crossing and $p_1 = (x_1; y_1)$ is the first negative point.

$$y(i) = x(i) * h(i); \tag{3}$$

$$\frac{dy(i)}{di} = x(i) * \frac{dh(i)}{di}; \tag{4}$$

$$i_{stripej} = x_0 - \frac{y_0(x_1 - x_0)}{y_1 - y_0}. \tag{5}$$

The filter is designed following the steps shown in Figure 12. Since the images are affected by impulsive noise, a Hamming window $h(i)$, as shown in Figure 12a, is chosen as the smoothing component. The derivate kernel d/di in Figure 12b is designed with a relatively large support to further mitigate the influence of local noise. Kernel $h'(i)$ designed to reduce noise and find the zero crossing is represented in Figure 12c. The filter size was set to 21 based on empirical observation of the laser stripe in the acquired images, whose maximum width was approximately 21 pixels, ensuring that the filter fully covered the stripe.

https://doi.org/10.3390/instruments10010007

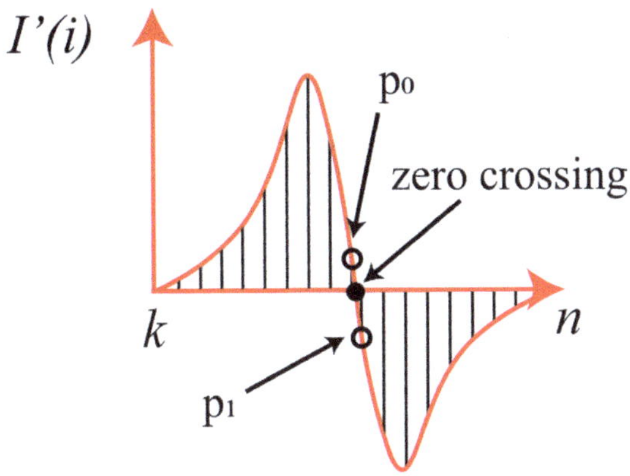

Figure 11. Graphical representation of FIR peak algorithm. Only for intensities above threshold is convolution computed.

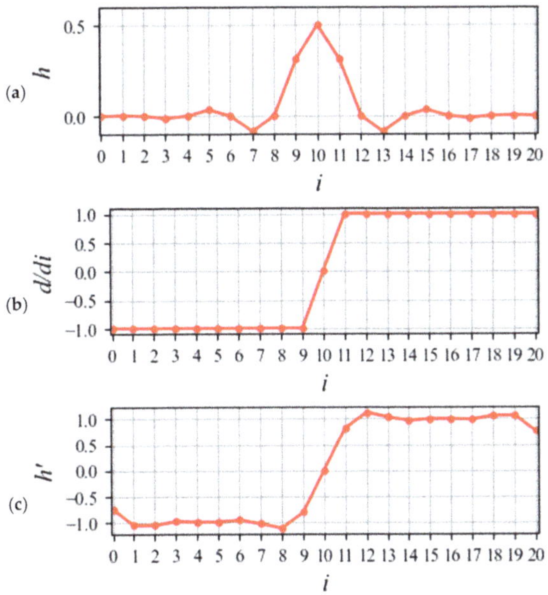

Figure 12. (**a**) Hamming filter with 21 parameters and cutoff of 0.5. (**b**) Derivate kernel. (**c**) FIR filter designed for this paper.

After finding the points, the laser stripe is obtained by linking each point with a scanning procedure: starting from the left, the point is joined to another if the distance between them is lower than the threshold.

All of this is performed in the sensor space with a plane of axes i and j. In order to transform sensor coordinates i and j into real-world coordinates x and z, the calibration functions $x(i,j), z(i,j)$ provided by the sensor are used. Calibration is not discussed in this paper; however, it consists of direct calibration using a reference geometrical artifact which is moved in the 3D space [17].

https://doi.org/10.3390/instruments10010007

3.5.3. Feature Extraction

In this paragraph, a feature extraction code is provided to evaluate gap and flush using the definition provided in Section 3.1 (Methodology). First, the parameters are set and Savitsky–Golay filtering [29] can be performed on the profile to extract the second-grade derivate Z'' that is calculated for both the left and right profile. As such, we can divide the profile into a straight line obtained considering second-derivative values below a threshold and an arc of a circle considering values above the same threshold. The hypothesis is that each profile is composed of at least one line and an arc of a circle. Lines and circles are evaluated for each profile according to the following pseudocode (Algorithm 1). A reference surface is chosen to evaluate gap and flush, as shown in Figure 2.

Algorithm 1. Finding features for one side (left or right).

Input: profile (x, z), linearSide, min_line_length l_{min}
Output: features → line, arc, edge point

(1) If linearSide = right: reverse (x, z).
(2) Set window_length = l_{min}; polyorder = 3; derivative order = 2. Set min_circle_length c_{min}. Set step st.
(3) Compute Savitsky–Golay second derivative on signal
(4) Build Boolean mask $B = \{|d_i| < \tau\}$. Split B into contiguous runs; drop runs of length $l < l_{min}$. Pick first run with length $l > l_{min}$ → line indices $\mathcal{L}$ (with optional trimming for very long lines)
(5) Let edge index i^* be last index in $\mathcal{L}$
(6) Circle search starting at $s = i^* + 1$.
(7) For windows $[s, e]$ of length $l \geq c_{min}$ with growth/shift rules:
    (a) Fit 3-point describing the arc → center C, radius r, length lr, angle θ.
    (b) If $r < r_{th}$ and $N_{outliers} \leq N_{max-outliers}$: mark as "arc" and try to extend e with st or shift window forward of st.
(8) Choose the best arc minimizing a cost function based on C, r, lr and θ.

3.6. Measurement System Analysis

This paper presents a benchmark to compare three different laser stripe extraction algorithms. The evaluation is based on two fundamental and equally important metrics: measurement uncertainty and processing time.

Processing time is a critical parameter, as the target G3F system is embedded and runs on a low-power SOM processing unit (based on the i.MX 8M family) designed for flexibility and low power consumption.

To robustly quantify measurement uncertainty, a GRR study was conducted, in accordance with standard MSA procedures, emulating a standard industrial process. This study is performed using the range method [23]. The range is calculated according to Equation (6):

$$R_{p,o} = \max(x_{o,p,t}) - \min(x_{o,p,t}), \tag{6}$$

where $x_{o,p,t}$ is the measure performed by operator o at point p in trial t. The GRR study breaks down the total variation in the measurement system into two primary components:

1. Repeatability uncertainty based on Equipment Variation (EV): the variation observed when a single operator measures the same part multiple times; this uncertainty is influenced by the uncertainty of the sensor and the uncertainty of the measurement method and is evaluated as

$$EV = \overline{R} \cdot K_1, \tag{7}$$

https://doi.org/10.3390/instruments10010007

where $\overline{R}$ is the average value of all ranges $R_{p,o}$ and K_1 is a statistical coefficient that depends on the number of trials.

2. Reproducibility uncertainty based on Appraiser Variation (AV): the variation that arises when different operators measure the same part; this uncertainty indicates the utility of training operators to obtain the same measurement method. It is calculated as in (8),

$$AV = \sqrt{\left(X_{diff} \cdot K_2\right)^2 - \left(\frac{EV^2}{P \cdot T}\right)}, \quad (8)$$

where P is the number of points measured, T is the number of trials, K_2 is a constant that depends on the number of operators, and X_{diff} is the range obtained considering only average values of measures for each operator, $\overline{x_o}$.

Finally, GRR is calculated as the combination of both contributions:

$$GRR = \sqrt{EV^2 + AV^2}. \quad (9)$$

4. Results

To evaluate the performance and robustness of the proposed gap and flush algorithms, a comprehensive MSA was carried out under different operating conditions. This chapter presents the outcomes of that assessment, comparing AV, EV, GRR, and processing time for the three algorithms considered. The analysis is performed both with and without the masking device mounted on the G3F system, allowing us to isolate the effects of hardware assistance and operator interaction on the measurement process. Each algorithm was executed on the same G3F SoM, based on the i.MX 8M Plus processor, in order to ensure comparable and representative processing time results for embedded applications.

The results reported in the following figures provide a clear overview of how each algorithm behaves in practical use and highlight the conditions under which performance degradation may occur.

Figure 13 shows the difference in performance, AV, EV, GRR, and processing time between the GC, FIR, and Steger algorithms while the prototyped front panel is used when measuring gap and flush. Figure 14 shows the same benchmark when the front panel is not mounted; here, it is possible to see a net decrease in performance for the FIR and Steger algorithms in gap analysis and for all the algorithms for flush analysis.

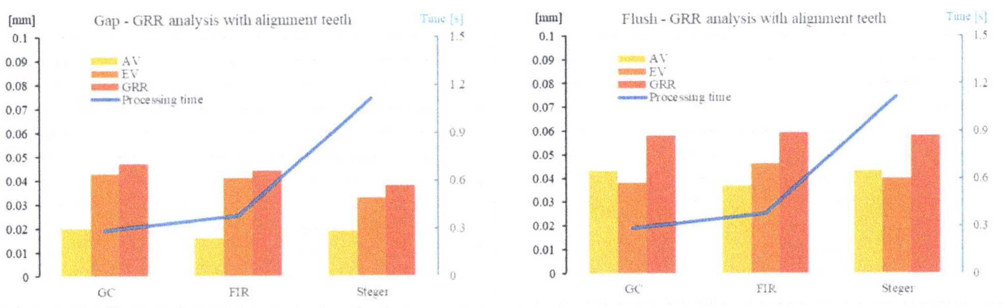

(**a**) (**b**)

Figure 13. Uncertainty and processing time benchmark using three different algorithms when front panel with alignment teeth is used to measure gap (**a**) and flush (**b**).

https://doi.org/10.3390/instruments10010007

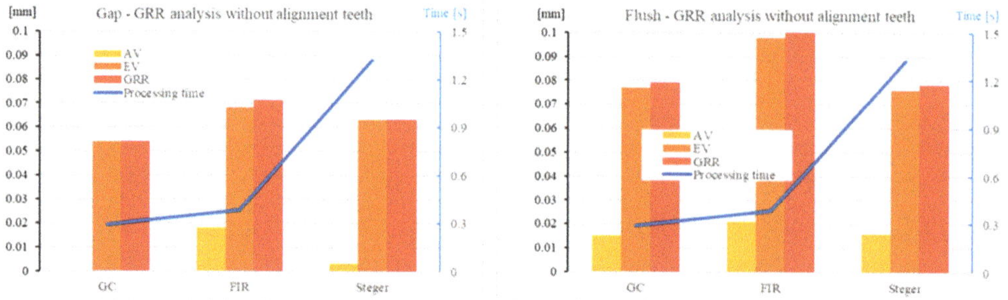

(**a**) (**b**)

Figure 14. Uncertainty and processing time benchmark using three different algorithms when front panel without alignment teeth is used to measure gap (**a**) and flush (**b**).

Quantitative values of GRR and processing time measurements are reported in Table 2. In this case, the selected algorithm is the GC, with a processing time of 0.29 s, which is significantly lower than those of the other methods: Steger, with processing times of 1.12 s and 1.32 s with and without alignment teeth, respectively, and FIR, with a processing time of 0.38 s.

Table 2. Results of processing time and GRR for each algorithm, considering images acquired with and without alignment teeth.

Algorithm	Processing Time with Alignment Teeth	Processing Time Without Alignment Teeth	Gap GRR with Alignment Teeth	Gap GRR Without Alignment Teeth	Flush GRR with Alignment Teeth	Flush GRR Without Alignment Teeth
GC	0.29 s	0.29 s	0.047 mm	0.054 mm	0.058 mm	0.079 mm
FIR	0.38 s	0.38 s	0.044 mm	0.071 mm	0.059 mm	0.1 mm
Steger	1.12 s	1.32 s	0.038 mm	0.063 mm	0.058 mm	0.078 mm

The GC also is more robust to sudden changes in operating conditions. The results shown also have other, more general practical consequences: the measurement system without alignment teeth has an internal uncertainty high enough to absorb the uncertainty caused by operator changes. We can define $\Delta EV = EV_{alignment} - EV$ as the difference between the EV calculated with alignment teeth and the EV calculated without alignment teeth. Similarly, ΔAV is calculated. FIR and Steger have $\Delta AV > \Delta EV$. Considering the measurement of gap, for the GC algorithm, $\Delta AV = 0.02$ mm and $\Delta EV = -0.01$ mm; for the FIR algorithm, $\Delta AV = 0$ mm and $\Delta EV = -0.03$ mm; for the Steger algorithm, $\Delta AV = 0.02$ mm and $\Delta EV = -0.03$ mm. Selecting the GC algorithm also reduces the need to redefine the measurement process considering that it has the lowest $|\Delta EV|$.

5. Conclusions and Limitations

This study set out to evaluate how algorithmic strategies and mechanical support features jointly affect the performance of a portable LLT system for gap and flush assessment in automotive applications. Through a dedicated MSA procedure, the work investigated whether more advanced extraction algorithms effectively enhance measurement reliability when the system is operated manually, and how mechanical alignment aids contribute to overall robustness. The following conclusions synthesize the main findings and outline the implications for the development of portable, operator-friendly metrology solutions.

The results show that the performance of the LLT measurement system is influenced not only by the choice of the extraction algorithm but also—often more significantly—by

https://doi.org/10.3390/instruments10010007

the mechanical conditions under which the measurement is performed. Although FIR and Steger algorithms incorporate more sophisticated processing techniques, their advantages decrease sharply when operator variability or suboptimal alignment conditions are present. The GC algorithm proved to be the most effective in this context: it consistently achieved lower processing times and maintained stable performance even when measurement conditions degraded.

The comparison between measurements taken with and without the alignment teeth on the front panel further highlights the dominant role of mechanical guidance. The absence of alignment features led to a marked increase in uncertainty—exceeding 30% for flush measurements—overshadowing the differences between algorithms. This finding demonstrates that, in portable metrology applications, mechanical support and ergonomic aids can have a stronger impact on repeatability and reproducibility than algorithmic complexity.

Future developments should therefore focus on improving both calibration and feature extraction. A promising direction is the integration of calibration procedures directly into the extraction stage, in a way that is coherent with the specific algorithm employed. Additionally, exploring mechanical modifications or user-assistance fixtures could further reduce operator influence and enhance the robustness of portable LLT systems.

Overall, this study demonstrates that achieving reliable, operator-assisted gap and flush measurements requires a balanced co-design of algorithms, calibration strategies, and mechanical support. Such an integrated approach aligns closely with the principles of Industry 5.0, where human-centred measurement systems must combine accuracy, repeatability, and usability to effectively support flexible and digitalized manufacturing environments.

Author Contributions: Conceptualization, G.G.P. and S.C.; investigation and methodology, G.G.P., S.C. and M.N.; software, G.G.P.; resources, C.C., M.N. and N.P.; writing—review and editing, G.G.P., C.C. and S.C.; supervision, S.C. and N.P. All authors have read and agreed to the published version of the manuscript.

Funding: This research received no external funding.

Data Availability Statement: The data are unavailable due to privacy restrictions.

Acknowledgments: This work was partially supported by PR Marche Fondo Sociale Europeo Plus 2021/2027 and the European project OPENZDM (Grant Agreement No. 101058673, call HORIZON-CL4-2021-TWINTRANSITION-01).

Conflicts of Interest: Matteo Nisi and Cristina Cristalli are employed by U-Sense IT Srl. The remaining authors declare no commercial or financial relationships that could be construed as a potential conflict of interest.

References

1. Bucci, I.; Fani, V.; Bandinelli, R. Towards Human-Centric Manufacturing: Exploring the Role of Human Digital Twins in Industry 5.0. *Sustainability* **2025**, *17*, 129. [CrossRef]
2. Nahavandi, S. Industry 5.0—A Human-Centric Solution. *Sustainability* **2019**, *11*, 4371. [CrossRef]
3. Buri, Z.; Kiss, J.T. Digitalisation in the Context of Industry 4.0 and Industry 5.0: A Bibliometric Literature Review and Visualisation. *Appl. Syst. Innov.* **2025**, *8*, 137. [CrossRef]
4. Stavropoulos, P. Digitization of Manufacturing Processes: From Sensing to Twining. *Technologies* **2022**, *10*, 98. [CrossRef]
5. Moona, G.; Singh, A.; Kumar, V.; Sharma, R.; Kumar, H. Dimensional Metrology: Underpinning the Automotive Sector in an Indelible Fashion. *MAPAN* **2024**, *39*, 931–941. [CrossRef]
6. D'Emilia, G.; Gaspari, A.; Natale, E. Accuracy Improvement of Measurement Management Systems: Requirements for Reliability of Data and Practical Examples. In Proceedings of the 2020 IEEE International Instrumentation and Measurement Technology Conference (I2MTC), Dubrovnik, Croatia, 25–28 May 2020; pp. 1–6.
7. Wang, H.; Li, M.; Wei, Q. Virtual Matching Measurement of Gap and Flush. In Proceedings of the 2021 14th International Symposium on Computational Intelligence and Design (ISCID), Hangzhou, China, 13–14 December 2021; pp. 192–197.

https://doi.org/10.3390/instruments10010007

8. Tran, T.-T.; Ha, C. Non-Contact Gap and Flush Measurement Using Monocular Structured Multi-Line Light Vision for Vehicle Assembly. *Int. J. Control Autom. Syst.* **2018**, *16*, 2432–2445. [CrossRef]
9. Minnetti, E.; Chiariotti, P.; Castellini, P.; Violini, L.; Garcia, G.; Vicente, H.; Paone, N. Smart Portable Laser Triangulation System for Assessing Gap and Flush in Car Body Assembly Line. In Proceedings of the 2019 II Workshop on Metrology for Industry 4.0 and IoT (MetroInd4.0&IoT), Naples, Italy, 4–6 June 2019; pp. 49–53.
10. Minnetti, E.; Chiariotti, P.; Paone, N.; Garcia, G.; Vicente, H.; Violini, L.; Castellini, P. A Smartphone Integrated Hand-Held Gap and Flush Measurement System for in Line Quality Control of Car Body Assembly. *Sensors* **2020**, *20*, 3300. [CrossRef] [PubMed]
11. Pham, L.H.; Tran, D.N.-N.; Rhie, C.H.; Jeon, J.W. A Mobile Vision-Based System for Gap and Flush Measuring between Planar Surfaces Using ArUco Markers. In Proceedings of the 2021 International Conference on Electronics, Information, and Communication (ICEIC), Jeju, Republic of Korea, 31 January–3 February 2021; pp. 1–4.
12. Kholkhujaev, J.; Maculotti, G.; Genta, G.; Galetto, M.; Inoyatkhodjaev, J. Non-Contact Articulated Robot-Integrated Gap and Flushness Measurement System for Automobile Assembly. *IEEE Access* **2022**, *10*, 86528–86541. [CrossRef]
13. Schröder, T.; Schwieger, V. Automated Gap and Flush Measurements between Car Parts Assisted by a Highly Flexible and Accurate Robot System: Fully Automated Gap Measurement in Agile Car Production. *J. Appl. Geod.* **2025**, *19*, 653–664. [CrossRef]
14. Pham, L.H.; Tran, D.N.-N.; Byun, J.Y.; Rhie, C.H.; Jeon, J.W. A Smartphone-Based Laser Measuring System for Gap and Flush Assessment in Car Body. *IEEE Trans. Ind. Electron.* **2021**, *68*, 6297–6307. [CrossRef]
15. D'Emilia, G.; Gaspari, A.; Iavicoli, C.; Natale, E. Measurement Uncertainty Estimation of Gap and Profile in the Automotive Sector. *J. Phys. Conf. Ser.* **2020**, *1589*, 012017. [CrossRef]
16. Paolini, G.G.; Nisi, M.; Casaccia, S.; Amicucci, F.; Cristalli, C.; Paone, N. Repeatability and Reproducibility Method of Hand-Handle Laser Triangulation Profilometer. In Proceedings of the 2025 IEEE International Workshop on Metrology for Industry 4.0 & IoT (MetroInd4.0 & IoT), Castelldefels, Spain, 1–3 July 2025; pp. 40–44.
17. Xu, X.; Fei, Z.; Yang, J.; Tan, Z.; Luo, M. Line Structured Light Calibration Method and Centerline Extraction: A Review. *Results Phys.* **2020**, *19*, 103637. [CrossRef]
18. Soucy, M.; Laurendeau, D.; Poussart, D.; Auclair, F. Behaviour of the Center of Gravity of a Reflected Gaussian Laser Spot near a Surface Reflectance Discontinuity. *Ind. Metrol.* **1990**, *1*, 261–274. [CrossRef]
19. Steger, C. An Unbiased Detector of Curvilinear Structures. *IEEE Trans. Pattern Anal. Mach. Intell.* **1998**, *20*, 113–125. [CrossRef]
20. Forest, J.; Salvi, J.; Cabruja, E.; Pous, C. Laser Stripe Peak Detector for 3D Scanners. A FIR Filter Approach. In Proceedings of the 17th International Conference on Pattern Recognition, 2004. ICPR 2004, Cambridge, UK, 26 August 2004; Volume 3, pp. 646–649.
21. Chen, X.; Fang, C.; Hu, A.; Ren, Y.; Zhang, M.; Cui, T. A Novel Reflective Interference Mitigation Model for Laser Stripe Extraction. *Measurement* **2024**, *237*, 115187. [CrossRef]
22. Bo, Q.; Hou, B.; Miao, Z.; Liu, H.; Lu, X.; Wang, Y. Laser Stripe Center Extraction Method Base on Hessian Matrix Improved by Stripe Width Precise Calculation. *Opt. Lasers Eng.* **2024**, *172*, 107896. [CrossRef]
23. Cepova, L.; Kovacikova, A.; Cep, R.; Klaput, P.; Mizera, O. Measurement System Analyses—Gauge Repeatability and Reproducibility Methods. *Meas. Sci. Rev.* **2018**, *18*, 20–27. [CrossRef]
24. *Computer Graphics: Principles and Practice*, 3rd ed.; Addison-Wesley: Upper Saddle River, NJ, USA, 2014; ISBN 978-0-321-39952-6.
25. Gonzalez, R.C.; Woods, R.E. *Digital Image Processing*, 4th ed.; Pearson Education: New York, NY, USA, 2018; ISBN 978-1-292-22304-9.
26. What Is Image Filtering in the Spatial Domain?—MATLAB & Simulink. Available online:https://it.mathworks.com/help/images/what-is-image-filtering-in-the-spatial-domain.html (accessed on 5 October 2025).
27. OpenCV: Image Thresholding. Available online: https://docs.opencv.org/4.x/d7/d4d/tutorial_py_thresholding.html (accessed on 19 January 2026).
28. Zhao, H.; Liu, X.; Wang, S.; Diao, K.; Luo, C. An Enhanced Centerline Extraction Algorithm for Complex Stripes in Linear Laser Scanning Measurement. *Precis. Eng.* **2024**, *91*, 199–211. [CrossRef]
29. Savitzky, A.; Golay, M.J.E. Smoothing and Differentiation of Data by Simplified Least Squares Procedures. *Anal. Chem.* **1964**, *36*, 1627–1639. [CrossRef]

https://doi.org/10.3390/instruments10010007

instruments

Article

Smartphone-Based Quantitative Measurement of Capillary Refill Time

Chiho Miyazawa [1,*], Masayoshi Shinozaki [2,3], Yayoi Miwa [4], Satoshi Karasawa [4], Taka-aki Nakada [4], Yukihiro Nomura [3] and Toshiya Nakaguchi [3,*]

1 Department of Medical Engineering, Graduate School of Science and Engineering, Chiba University, Chiba 263-8522, Japan
2 Augmented Living Technology Research Group (ALTec), Research Institute on Human and Societal Augmentation (RIHSA), National Institute of Advanced Industrial Science and Technology (AIST), Chiba 277-0882, Japan
3 Center for Frontier Medical Engineering, Chiba University, Chiba 263-8522, Japan
4 Department of Emergency and Critical Care Medicine, Chiba University Graduate School of Medicine, Chiba 260-8677, Japan
* Correspondence: miyazawa.chiho@chiba-u.jp (C.M.); nakaguchi@faculty.chiba-u.jp (T.N.)

Abstract

Capillary refill time (CRT) is widely used in pediatric and emergency medicine as an indicator of peripheral circulation. CRT is defined as the time required for the skin to return to its original color after external compression is applied and then released. In current clinical practice, however, CRT assessment remains qualitative and relies heavily on the magnitude and consistency of compression applied by the measurer, as well as on subjective visual color perception, which together result in limited measurement reliability. To improve measurement reliability, several quantitative CRT measurement devices have been developed. Nevertheless, these devices are dedicated specifically to CRT measurement, which limits their versatility and complicates clinical implementation. In this study, we developed a simple and quantitative CRT measurement method using a smartphone. Based on skin color changes captured by the rear camera, we proposed a method to assess the adequacy of the applied compression force and implemented an application to calculate CRT. In addition, we investigated an algorithm to reduce the influence of pulse waves observed in the post-release waveform, enabling more stable CRT estimation. Furthermore, a dedicated smartphone case was designed to immobilize the finger during measurement, thereby improving measurement reliability. The feasibility of the proposed method was evaluated by examining agreement with a previously developed CRT measurement device and by assessing intraexaminer reliability, confirming its effectiveness.

Keywords: capillary refill time; smartphone; RGB camera

1. Introduction

Capillary refill time (CRT) is used in pediatric and emergency medicine as an indicator for evaluating peripheral circulation. CRT is defined as the time required for blood to refill the capillaries after the chest or fingertips are compressed to expel capillary blood and the compression is subsequently released. A CRT exceeding 3 s may indicate shock associated with circulatory impairment [1]. Morin et al. reported that CRT accurately reflects microvascular blood flow and can serve as a predictor of mortality [2].

However, current CRT measurement relies on qualitative evaluation metrics dependent on the measurer's compression and color perception ability, posing a challenge due

https://doi.org/10.3390/instruments10010015

to its limitations. Moreover, visual assessment shows significant variability based on professional experience, with reports indicating particularly low reliability among less experienced healthcare professionals [3].

The development of quantitative CRT measurement devices has been pursued to improve CRT reliability [4,5], but strong demands from emergency responders and other users to minimize portable equipment have posed challenges to their widespread adoption.

Recently, several studies have reported blood flow measurement approaches using widely available smartphones [6,7]. Although CRT measurement using a smartphone (iPhone 8, Apple Inc., Cupertino, CA, USA) demonstrated good reproducibility, it remains dependent on the angle at which compression is applied with the finger, because compression is estimated based on changes in nail color [6].

In addition, a method has been proposed to estimate systolic and diastolic blood pressure by measuring photoplethysmography (PPG) signals using a smartphone [7]. In this approach, the PPG signal is acquired by placing the subject's index finger on the smartphone camera, processing RGB frames from the recorded video, and calculating the average intensity of the Green channel. Comparisons between blood pressure measured using a digital blood pressure monitor and values estimated by the proposed system showed high correlation for both systolic and diastolic blood pressure.

Therefore, we focused on smartphones as a measurement platform. If quantitative CRT measurement becomes feasible using smartphones, it could be performed simply by installing an application on devices that emergency responders already carry for communication. As a result, the need to transport dedicated equipment separately would be eliminated, substantially lowering barriers to system implementation. The objective of this study is to develop a method for measuring CRT using widely available mobile devices, such as smartphones.

2. Materials and Methods

2.1. Compression Force Quantification Based on Skin Color Change

Kawaguchi et al. [8] defined the optimal compression conditions for stable CRT measurement as compression exceeding 2 s with a force between 3 N and 7 N. To perform measurements under optimal compression conditions using mobile devices such as smartphones, it is necessary to evaluate whether the optimal compression force has been reached and whether compression exceeding 2 s at the optimal force is being applied, without relying on a force sensor. Therefore, we examined whether the achievement of optimal compression force could be evaluated from fingertip color changes without using a force sensor.

First, we describe a method for estimating the time at which the optimal compression force is reached (Figure 1). Fingertip color changes were observed using the output value of the Green component acquired from an RGB camera [4]. After compression begins, the Green value exhibits two stages, a rapid color change in the initial phase followed by a more gradual color change in the later phase [8]. The time from the start of compression until the appearance of the inflection point is defined as the inflection time, with this boundary treated as the inflection point. After the rapid color change, periodic color fluctuations caused by the pulse wave are observed, indicating that the increase in the Green value induced by compression has stabilized and that a sufficient compression state has been reached. The pulse wave onset time is defined as the first time point after the maximum increase time at which the first derivative of the Green value changes from positive to negative.

https://doi.org/10.3390/instruments10010015

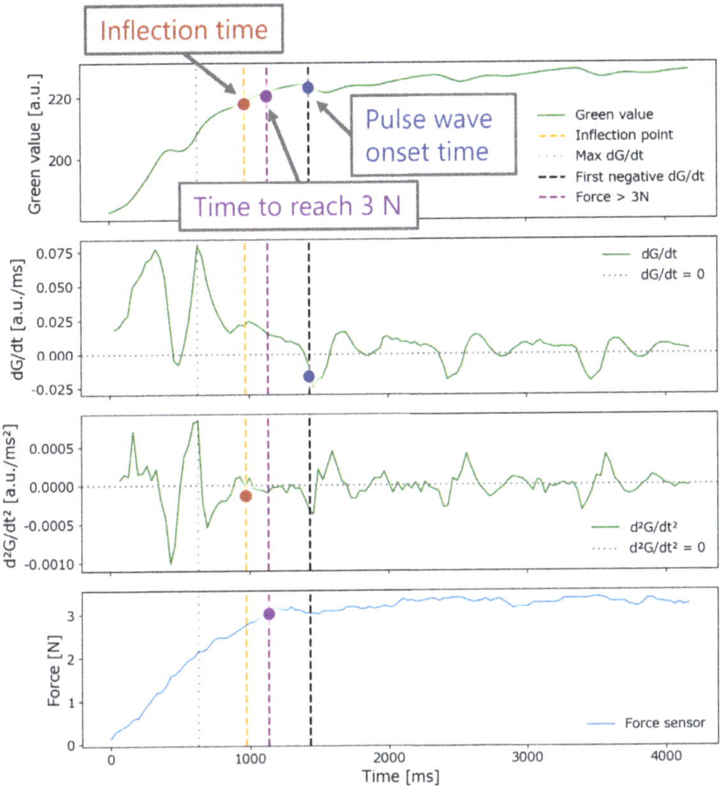

Figure 1. Calculation of optimal compression start time.

As a specific calculation procedure, the maximum increase time, defined as the time point at which the first derivative of the Green value reaches its maximum, is calculated within the first 3 s after compression begins. The inflection time is then defined as the first time point after the maximum increase time at which the second derivative changes from positive to negative. The pulse wave onset time is defined as the first time point after the maximum increase time, where the first derivative of the Green value changes from positive to negative. Due to noise or other factors, the pulse wave onset time may occasionally occur earlier than the inflection time. Therefore, the later of these two time points is defined as the optimal compression start time.

Next, we describe the method for evaluating whether compression was performed at the optimal level [9]. In CRT measurements, when compression is insufficient (1 N, 2 N), periodic fluctuations attributable to the pulse wave are observed in the Green value during compression (Figure 2).

Therefore, we propose the Pulse Presence Detection Index (PPDI) as a metric to quantify fluctuations in the Green value during compression. PPDI is used to evaluate the appropriateness of the applied compression force. First, the differential value of the Green signal during compression is calculated and squared. Next, the peak value of the squared differential signal within the negative region that is greater than or equal to the mean plus two times the standard deviation is calculated (Figure 3). PPDI is defined as 1000 times the mean value.

https://doi.org/10.3390/instruments10010015

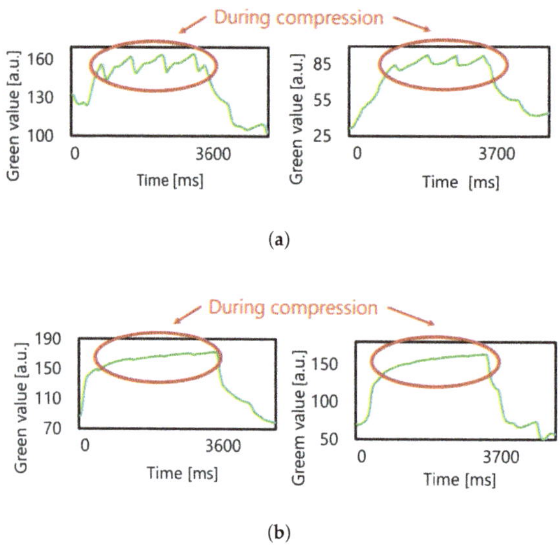

Figure 2. Recovery curve of Green value. (**a**) Example recovery curve obtained when compressing for 3 s below the optimal compression force. (**b**) Example recovery curve obtained when compressing for 3 s at the optimal compression force.

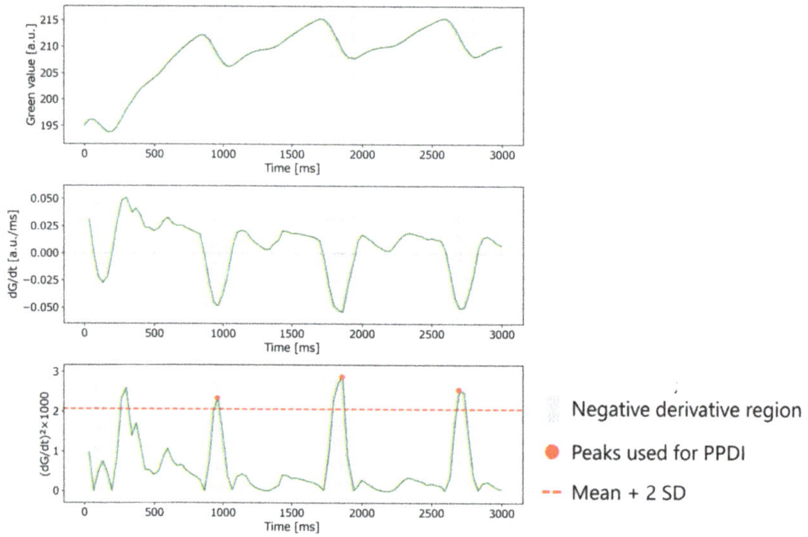

Figure 3. Calculation method for the Pulse Presence Detection Index. PPDI is defined as the average value of the peaks indicated by the red circle.

2.2. Determination of Optimal Compression Initiation Time

An experiment was conducted using six healthy adults in their 20s (5 males, 1 female), who alternated roles as subjects and measurers, to evaluate whether the optimal compression initiation time reached 3 N. All participants were Asian Skin Tone adults, and no extreme variations in skin tone were observed under the measurement lighting conditions. The experiment was conducted with approval from the Ethics Review Committee for Medical Research at Chiba University Graduate School (Approval Number: M10800).

https://doi.org/10.3390/instruments10010015

During the experiment, subjects lay supine and placed their left index finger against the rear camera and backlight of a smartphone (AQUOS sense2, Sharp Corporation, Sakai, Japan). Compression was applied to the subject's finger by the measurer using a force sensor mounted on the smartphone, within a range from the target compression force to the target compression force +0.5 N. The target compression forces were set to 3 N, 4 N, 5 N, 6 N, and 7 N. Compression force acquisition was performed using a calibrated capacitive force sensor (SingleTact CSU8-10N, SingleTact Ltd., Glasgow, UK). The measurement setup is shown in Figure 4. Three repeated measurements were performed at each compression force. Subjects and measurers were paired in groups of two and alternated roles after completing three repeated measurements.

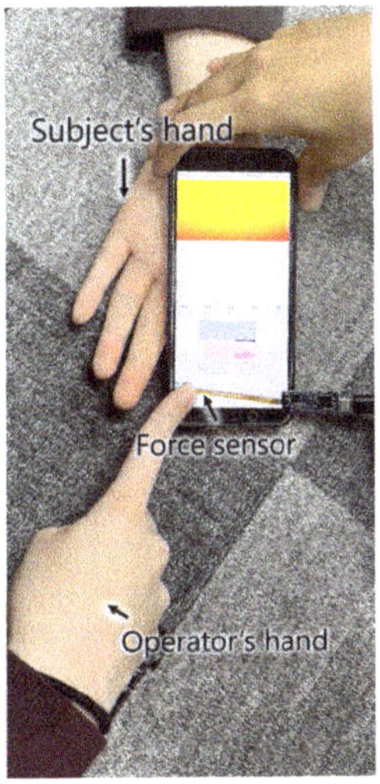

Figure 4. Experimental setup for determining the optimal compression initiation time.

2.3. *Evaluation of Compressive Force During Compression*

An experiment to evaluate compression force during compression application was conducted using six healthy adults in their 20 s (3 males, 3 females) as subjects. All participants were Asian Skin Tone adults, and no extreme variations in skin tone were observed under the measurement lighting conditions. One nonmedical professional in their 20 s served as the measurer. During the experiment, subjects lay supine and placed their left index finger against the rear camera and backlight of a smartphone (AQUOS sense2, Sharp Corporation, Sakai, Japan). The measurer applied compression to the subject's finger using a force sensor mounted on the smartphone, within a range from the target compression force to the target compression force +0.5 N. The target compression forces were set to 1 N, 2 N, 3 N, 4 N, 5 N, 6 N, and 7 N. Compression force acquisition was performed using a calibrated capacitive force sensor (SingleTact CSU8-10N, SingleTact Ltd., Glasgow, UK). Three repeated measurements were conducted at each compression force.

2.4. *Capillary Refill Time Measurement Application*

We developed a smartphone application that enables quantitative CRT measurement by positioning the backlight and rear camera of a smartphone (AQUOS sense2, Sharp Corporation, Sakai, Japan) against the subject's finger (Figure 5) and guiding compression and release through on-screen feedback. The application was developed and validated on Android version 10. Figure 5 illustrates the measurement overview in (a), the front view during measurement in (b), and the rear view during measurement in (c). The

https://doi.org/10.3390/instruments10010015

application instructs a three-second compression followed by a five-second release, and CRT is calculated using the five-second recording acquired after release. The average intensity of the Green channel is sampled at a frame rate of 30 fps.

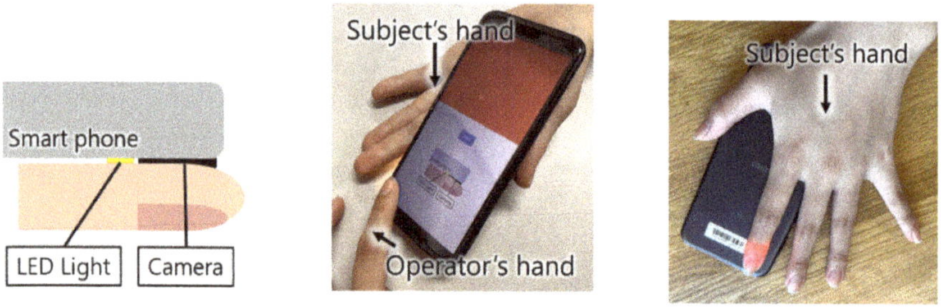

(a) (b) (c)

Figure 5. Measurement procedure using the application. (**a**) Overview diagram of smartphone-based measurement, (**b**) front view during measurement, and (**c**) rear view during measurement.

The application used in this study was developed using Flutter (version 3.27.3) and Dart (version 3.6.1). In addition, the camera settings were fixed at an ISO sensitivity of 700 and a shutter speed of 1/30 s, with white balance and focus distance also fixed. This configuration was used to prevent changes in imaging conditions during CRT measurements. However, when the Green value exceeded 250 during measurement, the signal was considered saturated, and the system prompted a remeasurement by reducing the ISO sensitivity by 100. Frames were acquired in YUV format with an eight-bit depth per channel and subsequently converted to RGB format to obtain the Green value.

2.5. Smartphone Case Development

To improve the reproducibility and stability of CRT measurements, we designed and fabricated a smartphone case (Figure 6). The case was manufactured using a 3D printer. The main body and securing band were fabricated from Thermoplastic polyurethane (TPU) material (flexibility 95A), whereas the fingertip positioning components were fabricated from Polylactic acid (PLA) material.

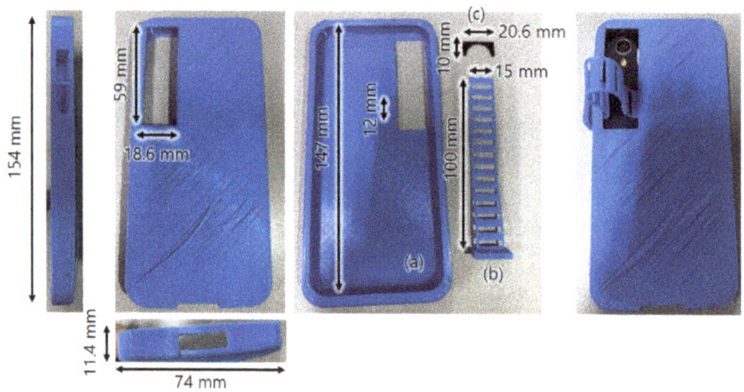

Figure 6. Smartphone case designed to improve the reproducibility and stability of CRT measurements. (a) Main case body, (b) band used to secure the finger, and (c) fingertip positioning part. The finger is inserted between the band and the smartphone, and the end of the band is then pulled to secure the finger in place.

https://doi.org/10.3390/instruments10010015

2.6. Capillary Refill Time Measurement Experiment

Next, to evaluate the validity of this measurement method, we assessed its consistency relative to CRT values measured using a portable measurement device [4] and evaluated intrarater reliability. A CRT measurement experiment was conducted with five healthy adults in their twenties (i.e., 4 males, 1 female) who had not participated in the experiments listed in Sections 2.2 or 2.3. All participants were Asian Skin Tone adults, and no extreme variations in skin tone were observed under the measurement lighting conditions. In addition, one nonmedical professional in their 20 s, consistent with Section 2.3, participated as the measurer. During the experiment, subjects lay supine and placed their left index finger against the rear camera and backlight of a smartphone (AQUOS sense2, Sharp Corporation, Sakai, Japan). The air conditioning was set to 26 °C. Before measurement, subjects rested quietly for five minutes to acclimate to the environment. Measurements were conducted under three conditions, use of a portable measurement device, smartphone measurement without a smartphone case, and smartphone measurement with a smartphone case attached. The measurement order was varied across subjects. Each subject underwent five repeated measurements, with a one-minute interval between successive measurements.

2.7. Data Preprocessing

The CRT calculation normalizes the maximum and minimum Green values after compression release, defining the time required to recover from 90% to 10% as $T(G_{10\%}) - T(G_{90\%})$ [8]. However, measurements obtained using smartphones exhibit pronounced periodic pulse wave fluctuations after decompression due to the weight of the smartphone, which makes detection of the 10% recovery point unstable. Therefore, the Green value at the time of decompression was used as the initial value, and CRT was calculated by fitting the postdecompression waveform using an exponential function model (Figure 7). The release time was defined as the moment at which the derivative of the Green value became negative.

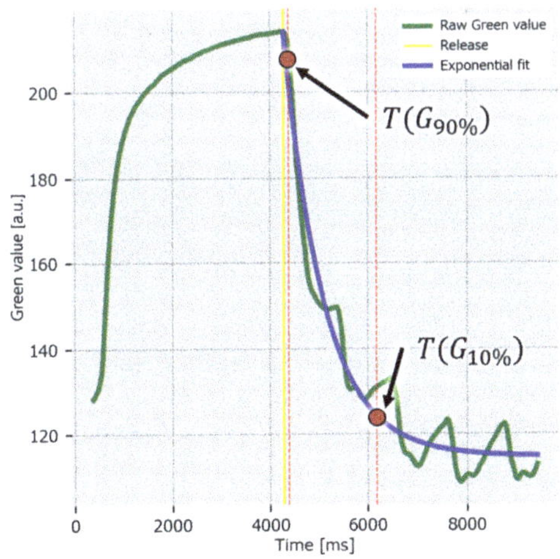

Figure 7. Approximation curve plotted using exponential fitting. The solid green line represents the observed Green values. The solid blue line represents the fitted curve obtained by applying an exponential model to the data after compression release. The red dashed lines indicate the 90% and 10% points defined by normalization based on the maximum and minimum values after compression release. The solid yellow line indicates the timing of compression release.

https://doi.org/10.3390/instruments10010015

2.8. Statistical Analysis

Statistical analysis was performed using R (version 4.4.3) and Python (version 3.11.9). R packages used included psych (version 2.5.6) and BlandAltmanLeh (version 0.3.1). Python libraries used included SciPy (version 1.15.2), statsmodels (version 0.14.6), Pingouin (version 0.5.5), and scikit-learn (version 1.7.2).

For all analyses, the threshold for statistical significance was set to $p < 0.05$.

3. Results

3.1. Optimal Compression Initiation Time

Figure 8a plots the calculated optimal compression start time against the time required to reach 3 N. All trials were included in the analysis. The dotted 45° line indicates agreement between the estimated and measured values, and the analysis focused on the region above this line. In 78 of 90 trials (86.7%), the optimal compression start time occurred after the time required to reach 3 N. Furthermore, as compression force increased, the proportion of cases in which 3 N had already been reached at the inflection time also increased (Figure 8b). These results demonstrate that when sufficient compression force is applied during compression and the pulse wave appearance time cannot be determined, the inflection time may be used as the optimal compression initiation time.

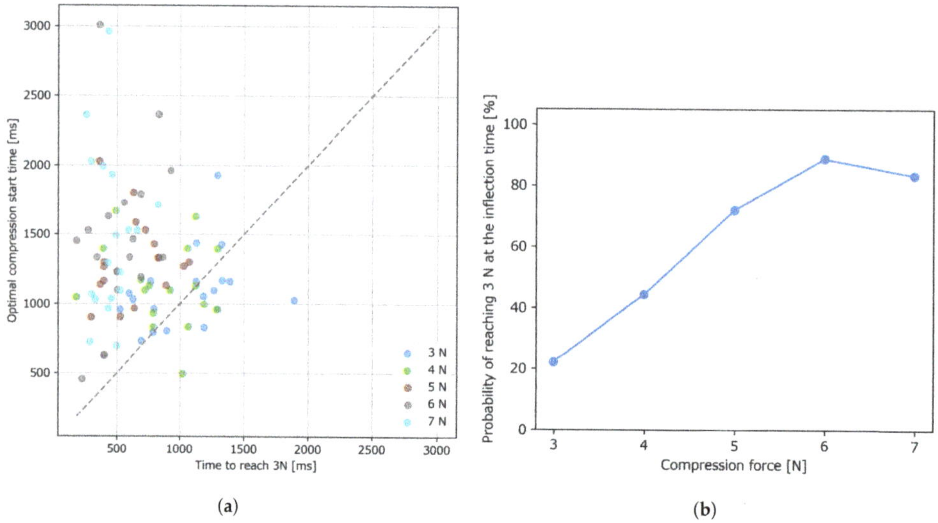

Figure 8. Results of experiments on optimal compression initiation time. (**a**) Relationship between optimal compression initiation time and the time required to reach 3 N. (**b**) Relationship between compression force and the percentage of trials in which 3 N was reached at the inflection point.

3.2. Compressive Force During Compression

Figure 9 shows the relationship between compression force and PPDI. All trials were included in the analysis. PPDI values measured below the optimal compression force were compared with those measured at the optimal compression force, revealing a significant difference (Mann–Whitney U test, U = 2264.0, $p < 0.001$). To evaluate generalization performance across subjects, Leave-One-Subject-Out cross-validation (LOSO-CV) was applied. In this analysis, compressions applied at 1–2 N below the optimal compression force were defined as insufficient compression and classified as positive, whereas compressions of 3 N or higher, corresponding to the optimal compression force, were classified as sufficient compression and labeled as negative. Receiver operating characteristic (ROC) analysis was

https://doi.org/10.3390/instruments10010015

performed on the training data to determine the optimal threshold, defined as the PPDI value that maximized Youden's index. The mean ± SD of the optimal PPDI threshold was 0.84 ± 0.60, and the area under the ROC curve (AUC) on the training data was 0.70 ± 0.03. The mean ± standard deviation of sensitivity and specificity on the validation data were 0.44 ± 0.36 and 0.88 ± 0.20, respectively. Therefore, PPDI of 0.84 or higher was determined to indicate insufficient compression force.

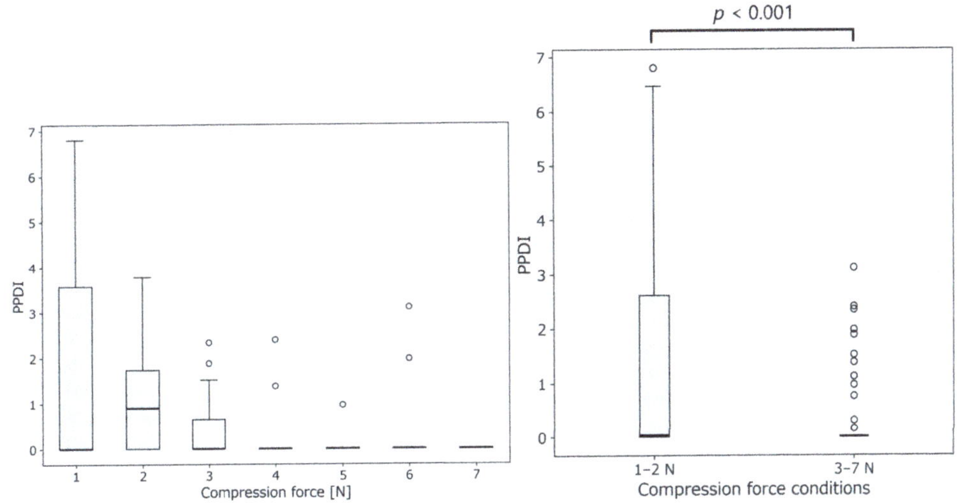

(a) (b)

Figure 9. Relationship between compression force and PPDI. (**a**) Relationship between each compression force and PPDI, and (**b**) relationship between PPDI for compression forces below the optimal level and at the optimal compression force.

3.3. Capillary Refill Time Measurement Experiment

The results obtained using the portable measurement device without a smartphone case and with a smartphone case are shown in Figure 10. All trials were included in the analysis.

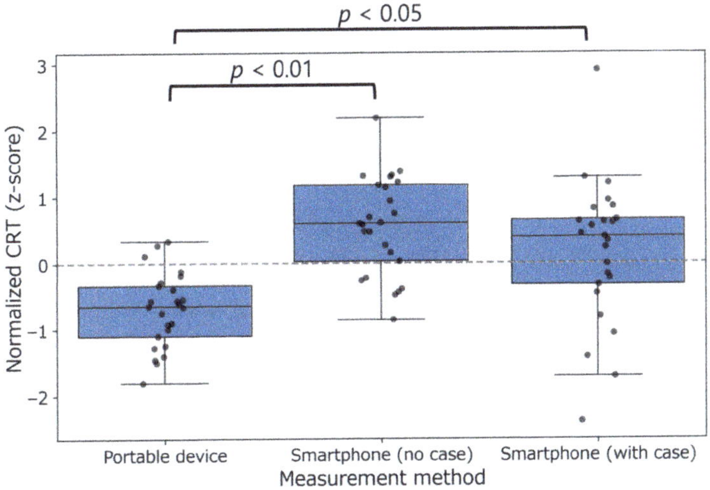

Figure 10. Box-and-whisker plots of CRT values normalized within subjects (z-scored) for each measurement method.

https://doi.org/10.3390/instruments10010015

Due to large interindividual differences among subjects, CRT was normalized (z-scored) within each subject as the deviation from the subject-specific mean, as shown in Equation (1), to evaluate relative changes as a preprocessing step.

$$CRT_{z,ij} = \frac{CRT_{ij} - \mu_i}{\sigma_i} \tag{1}$$

$CRT_{z,ij}$ denotes the normalized CRT for subject i and measurement number j, CRT_{ij} denotes the CRT for subject i and measurement number j, μ_i denotes the mean value of the CRT for subject i, and σ_i denotes the standard deviation of the CRT for subject i. The value of $CRT_{z,ij}$ is normalized to have a mean of zero and a variance of one for each subject.

Repeated measures ANOVA was used to assess differences across measurement methods and revealed a significant overall effect ($p = 0.0016$). Post hoc comparisons were performed using paired t-tests with multiple-comparison correction based on the Holm method. Significant differences were observed between the portable measurement device and the smartphone case present condition ($p\_adj = 0.035$), as well as between the portable measurement device and the smartphone case absent condition ($p\_adj = 0.0090$). In contrast, no significant difference was detected between the smartphone case present and smartphone case absent conditions ($p\_adj = 0.20$).

Figure 11 presents the results of Bland–Altman analysis [10] comparing measurements obtained using the smartphone and the portable measurement device. Analysis was conducted using the median of five repeated measurements. In Figure 11, the solid lines indicate the upper (LOA+) and lower (LOA−) limits of agreement, whereas the dashed lines represent the mean difference between measurement methods. For the results shown in Figure 11a, obtained using the portable measurement device without a smartphone case, LOA+ was 1.980 [95% CI 0.587 to 3.372] s and LOA− was −0.404 [95% CI −1.796 to 0.989] s. The mean difference between the two methods was 0.788 [95% CI 0.033 to 1.543] s, indicating the presence of a fixed error (one-sample t-test, $p = 0.0442$). No proportional error was detected (linear regression analysis, slope = 0.228, $p = 0.518$). For the results shown in Figure 11b, obtained using the portable measurement device with a smartphone case, LOA+ was 1.139 [95% CI 0.456 to 1.823] s and LOA− was −0.031 [95% CI −0.715 to 0.652] s. The mean difference between the two methods was 0.554 [95% CI 0.183 to 0.925] s, again indicating a fixed error (one-sample t-test, $p = 0.01427$). No proportional error was observed under this condition (linear regression analysis, slope = 0.122, $p = 0.488$).

We evaluated the agreement between smartphone measurements and those obtained using the portable measurement device using the intraclass correlation coefficient (ICC) [11]. Without a smartphone case, ICC(2,1) was 0.68 [95% CI −0.12 to 0.96]. With a smartphone case, ICC(2,1) was 0.84 [95% CI −0.06 to 0.98]. According to the criteria of Landis et al. [12], both conditions demonstrated fairly good agreement.

https://doi.org/10.3390/instruments10010015

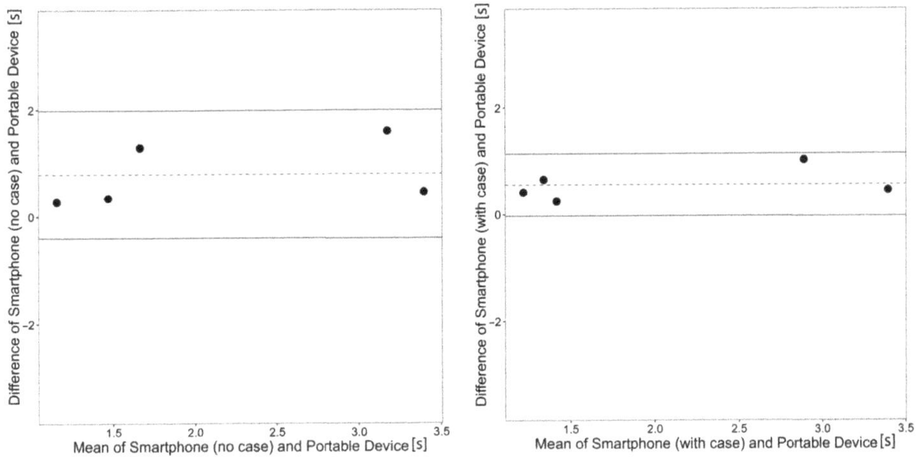

Figure 11. Evaluation of concordance between portable measurement device and smartphones in capillary refill time using Bland—Altman plots. (**a**) Results obtained using the portable measurement device without a smartphone case, and (**b**) results obtained using the portable measurement device with a smartphone case. The central blue dashed horizontal line indicates the mean difference (bias), and the upper and lower red solid horizontal lines represent the 95% limits of agreement (bias ± 1.96 SD).

In addition, intrarater reliability was evaluated. For the portable measurement device, ICC(1,1) was 0.90 [95% CI 0.71 to 0.99]. Without a smartphone case, ICC(1,1) was 0.87 [95% CI 0.63 to 0.98]. With the smartphone case, ICC(1,1) was 0.67 [95% CI 0.30 to 0.95]. Based on the criteria of Landis et al. [12], both measurement approaches showed fairly good repeatability.

4. Discussion

4.1. Optimal Compression Initiation Time

Figure 8a shows that at higher compression forces of 5 N or greater, the optimal compression force was reached by the optimal compression start time. In contrast, at lower compression forces of 3 N or 4 N, the optimal compression force was not consistently reached by the optimal compression start time. Figure 12 presents waveform examples from trials in which the optimal compression start time occurred before the time required to reach 3 N. When the compression force was low, the pulse wave was detected before the time to reach 3 N, indicating that this method has limited accuracy for estimating the time required to reach compression forces of 3 N or higher.

https://doi.org/10.3390/instruments10010015

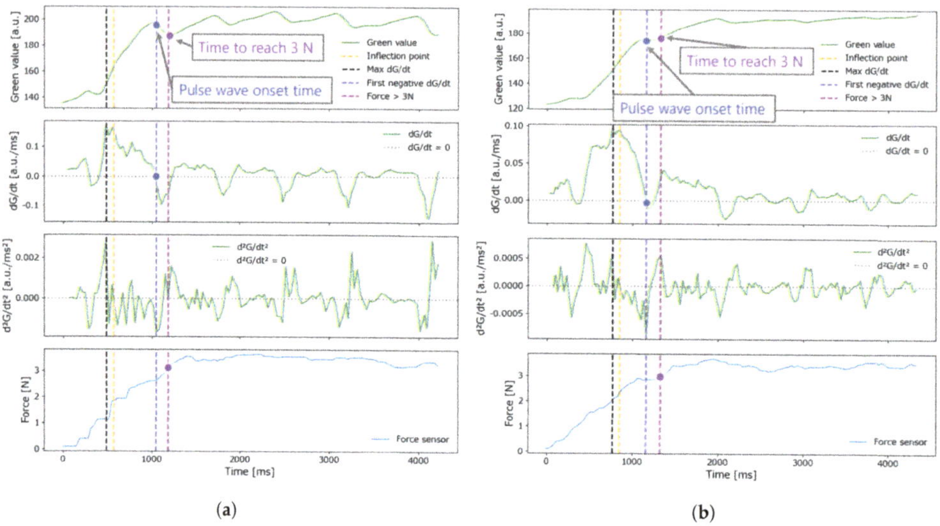

Figure 12. Waveforms from trials in which the optimal compression start time occurred before the time required to reach 3 N. The black vertical line indicates the time at which the first derivative of the G-value reached its maximum. The yellow vertical line indicates the inflection time. The blue vertical line indicates the pulse wave appearance time. The purple vertical line indicates the time at which 3 N was reached. (**a**) Subject 1's 3 N waveform and (**b**) Subject 3's 3 N waveform.

4.2. Compressive Force During Compression

The SD of the optimal PPDI threshold was relatively large at 0.60, indicating that PPDI exhibits substantial intersubject variability. Future studies should explore approaches for evaluating compression force during compression that are independent of subject-specific characteristics.

4.3. Capillary Refill Time Measurement Experiment

CRT values measured using smartphones tended to be longer relative to those measured using the portable measurement device. In smartphone-based measurements, because the smartphone rests directly on the subject's finger, the weight of the device continues to act on the finger even after compression is released. As a result, compression may not be fully relieved, which could affect the blood flow refill process. Similarly, Medic Hand reported by Shinozaki et al. [13] showed a fixed error relative to the portable measurement device due to insufficient finger release after compression. Therefore, it is necessary to consider a measurement approach that prevents the smartphone's weight from acting on the subject's finger or to reexamine the CRT calculation method. We are currently considering a design incorporating a simple lever or pivot mechanism to separate and eliminate the device's weight from its static force.

Figure 11 confirms that the difference between the median values obtained using the smartphone and the portable measurement device falls within the limits of agreement. This indicates consistency between the smartphone-based measurements and those obtained using the portable measurement device. In addition, the width of the limits of agreement was narrower when a smartphone case was used than when it was not. This finding indicates that use of a smartphone case reduces intersubject variability in measurement values, resulting in improved measurement accuracy. This effect is likely attributable to

https://doi.org/10.3390/instruments10010015

the bands and fingertip positioning components of the smartphone case, which reduce variability in finger placement.

Intraexaminer reliability was lower when using the smartphone case relative to measurements performed without the case. Figure 13 shows all waveforms obtained using the smartphone case for subject 3, who exhibited lower repeatability. Trial two shows a longer CRT relative to the other trials. In this trial, the pulse wave overlapped before the Green value had sufficiently recovered, which likely resulted in the prolonged CRT. A possible cause was inconsistent band tightness, suggesting difficulty in securing it with the same force each time. The developed band was designed to secure the fingertip. Securing the entire palm instead may allow hand fixation without obstructing fingertip blood flow. Future research should explore methods to secure the fingertip without obstructing its blood flow.

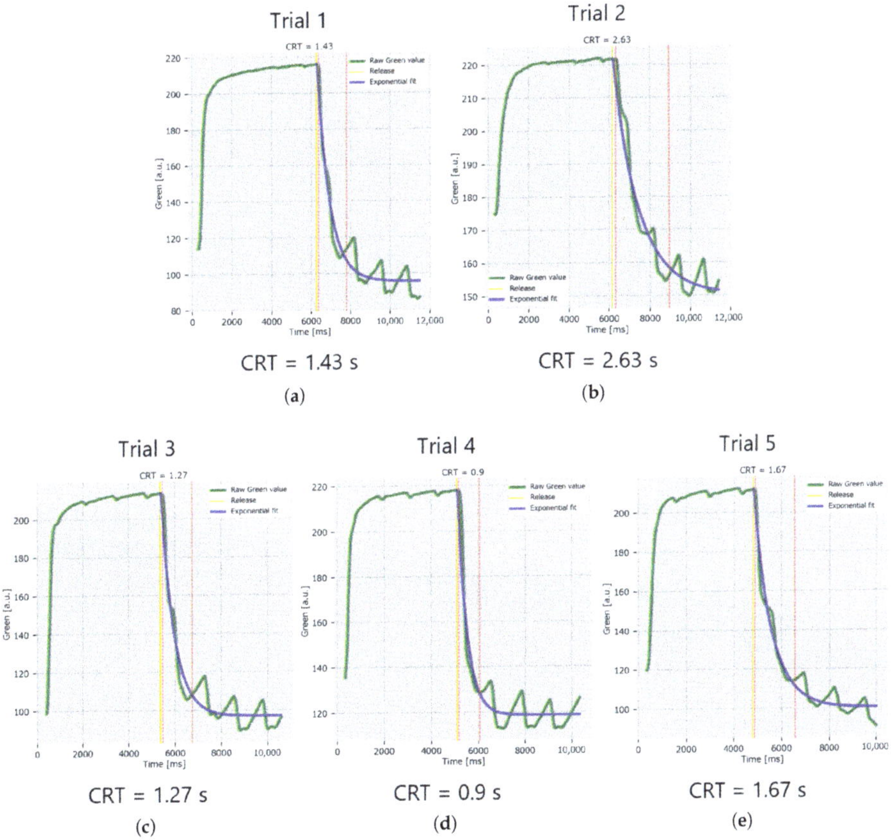

Figure 13. Recovery curves for all Green values obtained using the smartphone case for Subject 3. (**a**) Results from Trial 1, (**b**) results from Trial 2, (**c**) results from Trial 3, (**d**) results from Trial 4, and (**e**) results from Trial 5. The solid green line represents the observed Green values. The solid blue line represents the fitted curve obtained by applying an exponential model to the data after compression release. The red dashed lines indicate the 90% and 10% points defined by normalization based on the maximum and minimum values after compression release. The solid yellow line indicates the timing of compression release.

This study has several limitations. First, because a single examiner performed CRT measurements using both the portable measurement device and the smartphone, only intrarater reliability could be assessed. Future studies should evaluate inter-rater reliability

https://doi.org/10.3390/instruments10010015

by having multiple examiners assess the same subject. This is essential to ensure accurate CRT measurements by users other than the developer.

Second, the measurement interval in this study was set to one minute, although the optimal interval has not been fully investigated. Meyer et al. [14] conducted repeated CRT measurements at two minute and five minute intervals and found no difference between the two intervals. However, the shorter interval used in this study may have influenced repeated measurements.

Third, this study was conducted only on healthy individuals. To apply this technique in emergency medicine, similar experiments must be conducted on patients suffering from shock or admitted to the ICU, requiring more detailed investigation.

Fourth, Measurements using smartphones in this study were limited to the AQUOS sense2. Future work should consider expanding compatibility to other Android devices and iPhone devices.

Fifth, CRT is significantly affected by fingertip temperature, but no temperature compensation was applied in this study [15]. Implementing temperature compensation in future studies may enable more precise investigations.

Sixth, because this experiment was conducted with a limited number of subjects, further studies with a larger sample size are required, as well as evaluation of whether assessments performed by physicians align with the proposed methodology.

Seventh, the subjects in this study were limited to adults in their twenties. Because CRT is also used in pediatric populations, it is necessary to enable measurements in subjects with a wider range of finger sizes [16].

Eighth, comparative evaluation with other CRT calculation methods is also required. Descamps et al. [5] proposed a CRT calculation method based on the Lab color space. Meyer et al. [14] proposed a method based on TiVi values, in which CRT is calculated as the time required for skin color to return to its precompression state following decompression. Comparative analysis with these approaches should be addressed in future work.

5. Conclusions

This study developed a CRT measurement method using smartphones. We proposed an approach to evaluate whether the optimal compression force has been achieved and maintained, without using a force sensor, and assessed its ability to determine optimal compression force achievement in healthy subjects. Based on this evaluation, a threshold was established to identify whether compression was applied at the optimal level. In addition, we developed a CRT measurement application and a dedicated smartphone case. The feasibility of the proposed system was evaluated, and its effectiveness was confirmed by comparing measurement consistency and intraclass correlation coefficients with those obtained using a previously developed portable measurement device. Bland–Altman analysis demonstrated that differences between the median CRT values measured using the smartphone and the portable measurement device fell within the limits of agreement. Agreement between smartphone-based measurements and those obtained using the portable measurement device was further assessed using intraclass correlation analysis. Without a smartphone case, ICC(2,1) was 0.68 [95% CI −0.12 to 0.96], whereas with a smartphone case, ICC(2,1) increased to 0.84 [95% CI −0.06 to 0.98], indicating fairly good agreement. Intraexaminer reliability analysis showed an ICC(1,1) of 0.90 [95% CI 0.71 to 0.99] for the portable measurement device, an ICC(1,1) of 0.87 [95% CI 0.63 to 0.98] for smartphone measurements without a smartphone case, and an ICC(1,1) of 0.67 [95% CI 0.30 to 0.95] for smartphone measurements with a smartphone case, indicating fairly good repeatability. However, all subjects included in the present experiments were healthy

https://doi.org/10.3390/instruments10010015

individuals. Further investigation is required through similar experiments conducted in patients experiencing shock or those hospitalized in the ICU.

Author Contributions: Conceptualization, C.M., T.N. and T.-a.N.; methodology, C.M.; software, C.M.; formal analysis, C.M.; investigation, C.M.; writing—original draft preparation, C.M.; writing—review and editing, C.M., T.N., T.-a.N., Y.M., M.S., S.K. and Y.N.; visualization, C.M.; supervision, T.N. All authors have read and agreed to the published version of the manuscript.

Funding: This research received no external funding.

Institutional Review Board Statement: The Institutional Review Board of Chiba University Graduate School of Medicine approved this experimental study (Approval Number: M10800, Approval Data: 11 July 2025).

Informed Consent Statement: Informed consent was obtained from all subjects involved in the study.

Data Availability Statement: The datasets generated or analyzed during the current study are available from the corresponding author on reasonable request.

Conflicts of Interest: The authors declare no conflicts of interest.

References

1. World Health Organization. *Emergency Triage Assessment and Treatment (ETAT). Manual for Participants*; World Health Organization: Geneva, Switzerland, 2005.
2. Morin, A.; Missri, L.; Urbina, T.; Bonny, V.; Gasperment, M.; Bernier, J.; Baudel, J.L.; Kattan, E.; Maury, E.; Joffre, J.; et al. Relationship between skin microvascular blood flow and capillary refill time in critically ill patients. *Crit. Care* **2025**, *29*, 57. [CrossRef] [PubMed]
3. Shinozaki, K.; Jacobson, L.S.; Saeki, K.; Kobayashi, N.; Weisner, S.; Falotico, J.M.; Li, T.; Kim, J.; Lampe, J.W.; Becker, L.B. The standardized method and clinical experience may improve the reliability of visually assessed capillary refill time. *Am. J. Emerg. Med.* **2021**, *44*, 284–290. [CrossRef] [PubMed]
4. Shinozaki, M.; Shimizu, R.; Saito, D.; Nakada, T.; Nakaguchi, T. Portable measurement device to quantitatively measure capillary refilling time. *Artif. Life Robot.* **2022**, *27*, 48–57. [CrossRef]
5. Descamps, A.; Jacquet-Lagrèze, M.; Aussal, T.; Fellahi, J.-L.; Ruste, M. DiCART device to measure capillary refill time: A validation study in patients with acute circulatory failure. *J. Clin. Monit. Comput.* **2025**, *39*, 831–840. [CrossRef] [PubMed]
6. Tsunomura, M.; Nagasawa, T.; Takahashi, R.; de Souza, R.P.; Ogawa, K.; Tsumura, N.; Cardoso, G.C. Smartphone measurement of skin capillary refill time. In *OSA Imaging and Applied Optics Congress*; Optica Publishing Group: Washington, DC, USA, 2021; p. ITu7A.5.
7. Tabei, F.; Gresham, J.M.; Askarian, B.; Jung, K.; Chong, J.W. Cuff-less blood pressure monitoring system using smartphones. *IEEE Access* **2020**, *8*, 11534–11545. [CrossRef]
8. Kawaguchi, R.; Nakada, T.; Oshima, T.; Shinozaki, M.; Nakaguchi, T.; Haneishi, H.; Oda, S. Optimal pressing strength and time for capillary refilling time. *Crit. Care* **2019**, *23*, 4. [CrossRef] [PubMed]
9. Ma, Z.Y.; Sun, S.; Wu, S.C.; Lin, L.; Chen, Y.X.; Zhao, D.; Morgan, S.P. Pilot study on optimizing pressure for standardized capillary refill time measurement. *Heliyon* **2024**, *10*, e35716. [CrossRef] [PubMed]
10. Bland, J.M.; Altman, D.G. Statistical methods for assessing agreement between two methods of clinical measurement. *Lancet* **1986**, *327*, 307–310. [CrossRef]
11. Wastell, D.G.; Barker, G.R. Intraclass correlations: A two-facet case study and some comments on the concept of reliability. *Bull. Psychon Soc.* **1988**, *26*, 583–586. [CrossRef]
12. Landis, J.R.; Koch, G.G. The measurement of observer agreement for categorical data. *Biometrics* **1977**, *33*, 159–174. [CrossRef] [PubMed]
13. Shinozaki, M.; Saito, D.; Nakada, T.; Nomura, Y.; Nakaguchi, T. Feasibility study of wearable capillary refill time measurement device. *Artif. Life Robot.* **2024**, *29*, 334–339. [CrossRef]
14. Meyer, F.; Henricson, J.; Anderson, C.D.; Wilhelms, D.B. The effect of repeated capillary refill tests on the cutaneous microcirculation. *J. Biophotonics* **2024**, *17*, e202400098. [CrossRef] [PubMed]

https://doi.org/10.3390/instruments10010015

15. Shinozaki, K.; Capilupi, M.J.; Saeki, K.; Hirahara, H.; Horie, K.; Kobayashi, N.; Weisner, S.; Kim, J.; Lampe, J.W.; Becker, L.B. Low temperature increases capillary blood refill time following mechanical fingertip compression of healthy volunteers: Prospective cohort study. *J. Clin. Monit. Comput.* **2019**, *33*, 259–267. [CrossRef] [PubMed]
16. Fernández-Sarmiento, J.; Lamprea, S.; Barrera, S.; Acevedo, L.; Duque, C.; Trujillo, M.; Aguirre, V.; Jimenez, C. The association between prolonged capillary refill time and microcirculation changes in children with sepsis. *BMC Pediatr.* **2024**, *24*, 68. [CrossRef] [PubMed]

https://doi.org/10.3390/instruments10010015

instruments

MDPI

Article

Wireless Instrumented Ankle Foot Orthosis (AFO) for Gait Cycle Monitoring

Soufiane Mahraoui * and Mauro Serpelloni

Department of Information Engineering, University of Brescia, 25123 Brescia, Italy; mauro.serpelloni@unibs.it
* Correspondence: soufiane.mahraoui@unibs.it

Abstract

Ankle–foot orthoses (AFOs) are widely used in the rehabilitation of patients with neurological or musculoskeletal disorders. However, treatment outcomes may be influenced by incorrect use of the device or by inappropriate orthosis selection. Since many types of AFOs are available, differing in materials, stiffness, and geometry, an objective evaluation tool can support clinical decision-making. This work presents the design, development, and characterization of an instrumented AFO able to quantify relevant gait parameters in an objective way. The proposed device integrates three measurement modalities in a compact wearable structure. Two longitudinal strain gauges estimate ankle plantar- and dorsiflexion angles. Two force-sensitive elements detect foot–ground contact and allow identification of stance and swing phases of the gait cycle. A single inertial measurement unit (IMU) is used to measure lateral shank inclination. The strain-gauge-based angle estimation was validated against a gold-standard motion capture system, achieving a root mean square error of approximately 1.6 degrees and showing higher accuracy than the IMU for plantar/dorsiflexion measurement, while maintaining a simple electronic architecture. The force sensors were validated using a force platform and demonstrated reliable detection of loading and unloading events. Monitoring lateral inclination through the single IMU provides additional information related to balance and potential fall risk. Data are transmitted via Bluetooth Low Energy (BLE) to a custom Python-based application for real-time visualization and recording. Overall, the results validate the electronic instrumentation and demonstrate reliable system performance, indicating that the proposed instrumented AFO represents a promising platform for objective gait assessment and future clinical applications.

Keywords: gait monitoring; ankle–foot orthosis; instrumented AFO; ankle angle estimation; strain gauges; ground contact detection; single IMU; rehabilitation assessment

1. Introduction

Gait analysis is essential for understanding and treating walking disorders. Modern sensing technologies enable objective and quantitative monitoring of gait, supporting rehabilitation and clinical decision making.

The gait cycle is the sequence of movements between two consecutive contacts of the same foot with the ground [1]. It consists of a stance phase, when the foot is in contact with the ground (about 60% of the cycle), and a swing phase, when the foot moves in the air (about 40%).

Two key parameters for gait analysis are ankle plantar/dorsiflexion and lateral shank tilt. Ankle dorsiflexion is the upward movement of the foot toward the shin, while plantarflexion is the downward movement. Normal ankle motion [2] during gait ranges approximately from $-15°$ (dorsiflexion) to $15°$ (plantarflexion).

Wireless communication is essential for improving system portability and usability. Low-power protocols enable reliable data transmission with extended operating time, and many systems adopt IoT architectures with cloud-based processing [3]. However, these solutions may introduce latency, higher energy consumption, and dependence on network connectivity.

Lateral shank tilt describes the inclination of the lower leg in the medial–lateral direction. Medial means toward the body midline, while lateral means away from it. Monitoring this angle helps evaluate balance and load distribution; excessive tilt can indicate instability and a higher risk of falls [4].

The integration of wearable sensing systems into assistive devices, such as ankle-foot orthoses (AFOs) or walking aids [5–9], enables clinicians to obtain objective and quantitative data on an individual's gait and mobility patterns directly within clinical settings. By continuously or periodically monitoring temporal, spatial, and kinematic parameters of the gait cycle, these systems support the functional assessment of walking performance and facilitate the identification of deviations from normative patterns. Such data-driven insights are essential for informing clinical decision-making and refining personalized rehabilitative strategies and therapeutic adjustments [10–12].

Foot drop is a common gait pathology characterized by reduced dorsiflexion due to weakness or paralysis of the dorsiflexor muscles. This deficit leads to excessive plantarflexion during the swing phase, increasing the risk of toe dragging and falls. A Posterior Leaf Spring AFO (PLS-AFO) assists dorsiflexion during swing while limiting excessive plantarflexion during stance, improving foot clearance and gait safety [13].

In recent years, several instrumented assistive devices have been developed to monitor specific gait parameters. These include ankle or crutch kinematics [14,15], Ground Reaction Force (GRF) [16], orthosis deformation [17], and muscle activity [18]. For example, Hamid et al. [19] proposed an active AFO system incorporating a force sensor placed under the foot to detect gait phases, enabling assistance for patients affected by foot drop [20]. Rescio et al. [21] developed an instrumented insole capable of measuring foot temperature, which has been shown to correlate with diabetic complications such as foot ulcers. Strain gauges have been widely employed to investigate orthosis deformation. Svensson et al. [17] used a Wheatstone bridge configuration to measure deformation during uphill and downhill walking, demonstrating higher deformation during ascent. Similarly, Tanino et al. [22] implemented strain gauge bridges to estimate joint torque during gait. These approaches exploit the direct relationship between structural deformation and joint motion.

Recent advances in printed electronics have enabled the integration of sensing elements into medical devices [23–26], particularly orthotic systems [27,28]. These technologies allow sensors to be directly embedded onto orthotic structures, improving wearability and customization. Nevertheless, most existing systems either focus on isolated parameters or rely on multiple IMUs for joint kinematics estimation. Therefore, reducing hardware complexity while preserving measurement capability remains an open challenge.

In parallel, inertial sensors are commonly used to identify gait events and estimate spatio-temporal parameters. Aminian et al. [29] combined gyroscopes with footswitch sensors to detect initial contact and toe-off, while Betz et al. [30] investigated ankle kinematics using IMUs integrated into orthoses. However, accurate ankle angle estimation typically requires a multi-IMU configuration, with one sensor on the foot and another on the shank, to reconstruct joint kinematics from relative segment orientation [31].

Although effective, this approach increases system complexity and calibration effort [32], as precise alignment and synchronization between sensors are required [33]. It

https://doi.org/10.3390/instruments10020023

is also sensitive to positioning errors, since small variations in sensor placement after reattachment can affect measurement accuracy [34]. In addition, advanced sensor fusion methods increase computational demand, limiting real-time or low-power applications. The use of multiple active sensors further raises power consumption and reduces battery life [35], while wearable units on both foot and shank may negatively impact user comfort and compliance [36].

To address these limitations, this study proposes an alternative approach based on the combination of strain gauges and a single IMU. Building upon our preliminary work [37], strain gauges embedded in the AFO structure measure deformation induced by ankle motion, acting as a direct proxy for joint angle. Unlike multi-IMU systems, this method does not rely on inter-segment kinematic reconstruction. A single IMU, positioned on the shank, provides complementary information for gait phase detection and orientation.

The system integrates strain gauges for deformation monitoring, force sensors for ground contact detection, and one IMU for gait phase estimation. Data are transmitted via Bluetooth Low Energy (BLE) to a custom Python-based interface, avoiding cloud dependency. A dedicated PCB and a 3D-printed support structure ensure seamless integration within the orthosis.

In summary, the proposed approach replaces the conventional multi-IMU configuration with a deformation-based sensing strategy supported by a single IMU. Compared to traditional systems, it reduces hardware complexity by eliminating the foot-mounted sensor, simplifies calibration through a direct strain–angle relationship, and improves robustness to sensor repositioning. At the same time, it lowers power consumption and enhances user comfort by minimizing wearable components. Despite this simplification, the approach maintains reliable ankle angle estimation, offering a practical trade-off between measurement performance and system complexity.

The objective of this study is to validate the proposed system through experiments on healthy subjects, using an optical motion capture system as a reference. The main contribution is to demonstrate that orthosis deformation sensing, combined with inertial data, enables accurate ankle angle estimation while significantly reducing instrumentation requirements.

2. Wireless AFO Description

The goal of this study is to develop an instrumented Posterior Leaf Spring Ankle-Foot Orthosis (PLS-AFO)—as shown in Figure 1—capable of monitoring multiple biomechanical parameters during gait. Specifically, the system aims to measure the deformation of the orthosis itself, which allows for the calculation of plantarflexion and dorsiflexion angles at the ankle joint. Additionally, the device tracks the phases of gait, distinguishing between stance and swing, and captures the acceleration and spatial positioning of the lower leg. An important feature is the monitoring of the lateral shank tilt angle, which provides valuable information on balance and stability during walking.

To make these measurements, the instrumented PLS-AFO integrates different sensors. Strain gauges are placed on the orthosis to measure its deformation, which is related to ankle plantarflexion and dorsiflexion. In a typical setup, measuring these angles requires one IMU placed on the shank and another on the foot to estimate the relative segment orientation. In contrast, the proposed solution allows the ankle angle to be estimated without using two IMUs. Therefore, only one IMU is needed, reducing system complexity, calibration effort, power consumption, and improving user comfort.

Two force sensors are used to detect key gait events, specifically heel strike and toe-off (toe strike), allowing identification of gait phases. Furthermore, an inertial measurement unit (IMU) is attached to the leg to monitor acceleration and position in real time. In

https://doi.org/10.3390/instruments10020023

particular, the tilt angle can be obtained indirectly using this acceleration and position, allowing the assessment of lateral shank tilt during the gait cycle.

All sensor signals are collected and processed by a microcontroller embedded within the orthosis. The data are then transmitted wirelessly via Bluetooth Low Energy (BLE) to a personal computer. On the computer, a dedicated graphical user interface (GUI) displays the real-time data to clinicians, facilitating immediate analysis and feedback. Additionally, the system saves the acquired data in a text (.csv) file for further offline processing and study.

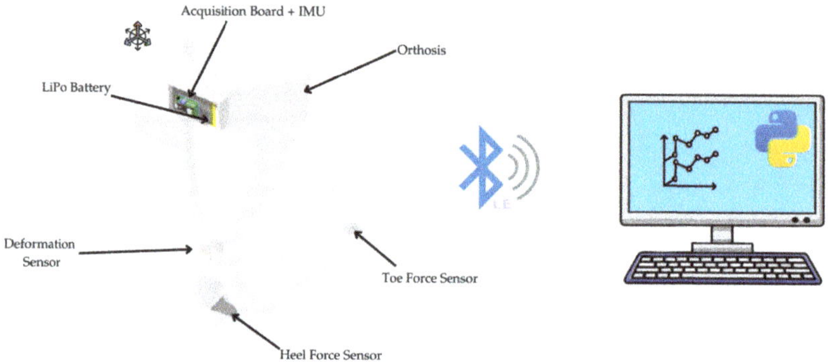

Figure 1. Instrumented AFO.

2.1. Force Sensors

Force sensors are placed at the bottom of the orthosis to detect key phases of the gait cycle, namely heel strike and toe-off. Their location is selected to maximize the accuracy of gait event detection. The adopted sensors are resistive force sensors, which require proper signal conditioning, as shown in Figure 2. To this end, a non-inverting amplifier configuration is employed. An offset voltage of approximately 0.5 V is applied at the amplifier input to prevent the output from reaching negative supply levels, thereby ensuring that the signal remains within the admissible voltage range. As pressure is applied to the sensor, the output voltage increases proportionally, approaching the supply voltage with increasing force. The gain resistor (R_G) is set to 892 Ω, such that the amplifier output saturates at an applied force of approximately 40 N. This threshold is defined according to the criterion that ground contact is detected when the applied force exceeds 5% of the subject's body weight [38,39]. A characterization of the force sensor is presented in the Section to determine its resistance value at 40 N. This configuration effectively maps the analog sensor output into a binary signal, enabling reliable detection of ground contact events. Each sensor is implemented using the same configuration.

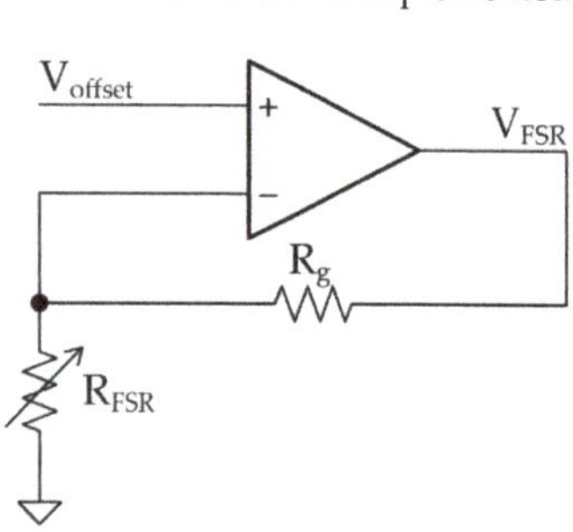

Figure 2. Force Sensor—Circuit.

https://doi.org/10.3390/instruments10020023

2.2. Strain Gauges Sensors

To measure the deformation of the orthosis, two 120 Ω strain gauges are mounted on the structure in a longitudinal position, as shown in Figure 3 within a red square. One sensor is placed on the outer side of the orthosis, while the other is positioned on the inner side. This configuration allows the sensors to experience opposite mechanical effects during deformation: when one strain gauge is subjected to compression, the other is subjected to tension. In this study, a full Wheatstone bridge configuration is also tested by adding two additional strain gauges placed orthogonally to the first pair, in a transversal position. This configuration is evaluated to determine whether the sensitivity and accuracy of the bridge can be enhanced.

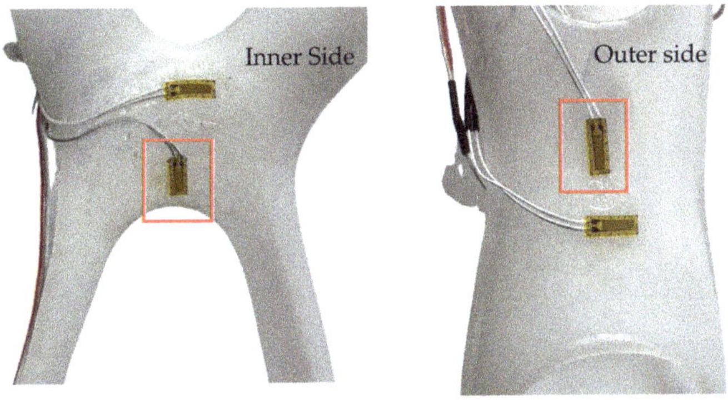

Figure 3. Strain gauges positioning in a full Wheatstone bridge; inside the red square, the longitudinal strain gauges corresponding to a half Wheatstone bridge.

A half Wheatstone bridge (Figure 4) is implemented using two commercial resistors with a nominal value of 120 Ω, matching the nominal resistance of the strain gauges under no-load conditions. In this configuration, the bridge is initially balanced when no external forces are applied. When the strain gauges are deformed by ankle movement, their resistance changes, generating a differential voltage at the bridge output. The voltage is positive during plantarflexion and negative during dorsiflexion, reflecting the direction of the applied force.

The differential signal from the bridge is measured using an instrumentation amplifier (INA) configured with a gain of 500, obtained by selecting a gain resistor of 124 Ω according to the standard INA gain formula. A reference voltage of 1.5 V is applied to the INA to properly shift the signal within the acquisition range, since the Wheatstone bridge produces both positive and negative voltages. The gain and reference values were chosen following the standard INA design equations.

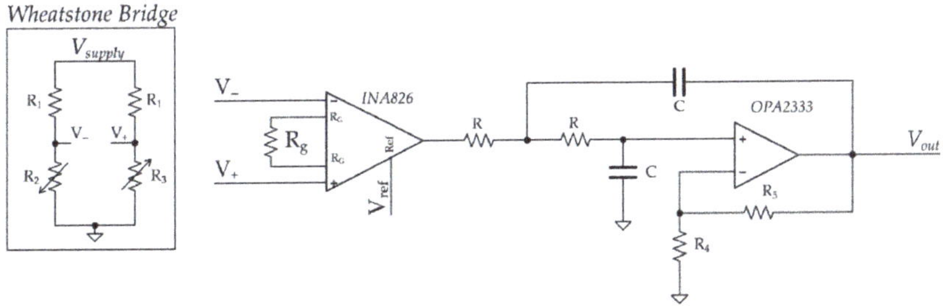

Figure 4. Strain Gauges—Conditioning circuit.

https://doi.org/10.3390/instruments10020023

The amplified signal is then processed by a Sallen–Key low-pass filter, which also provides additional amplification with a gain of 1.5. The filter has a cutoff frequency of 25 Hz, chosen to satisfy the Nyquist criterion given the 66 Hz sampling rate of the system. The resistor and capacitor values of the filter were calculated according to standard Sallen–Key formulas. This configuration allows proper scaling of the signal and full use of the 0–3.3 V input range of the acquisition system.

2.3. Hardware System

The strain gauges and force sensors, as described in the previous section, are equipped with an electronic conditioning circuit to properly amplify and filter their signals. The conditioned outputs are then acquired by a 12-bit analog-to-digital converter (ADC), which provides a resolution of approximately 1 mV per step. The ADC is integrated within an Arduino Nano 33 BLE Sense Rev2, which also contains the inertial measurement unit (IMU) for accelerometer and gyroscope measurements, as well as a Bluetooth Low Energy (BLE) module for wireless data transmission. The entire system is powered by a 3.7 V LiPo battery. The supply voltage is stepped down to 3.3 V using a low-dropout (LDO) buck converter, which provides a stable voltage directly to the Arduino microcontroller. This direct supply bypasses the onboard protection circuitry of the Arduino, which is necessary because the Arduino Nano 33 BLE Sense Rev2 can only operate reliably with a 3.3 V supply voltage in this configuration. The power management system also includes a battery charging module (TP4056) via USB-C, which ensures safe charging of the LiPo battery. The module features a red LED indicator, which lights up when the battery is charging. This charging module is necessary to protect the LiPo battery and prevent safety issues related to overcharging or short circuits. Thanks to the LDO buck converter, the microcontroller receives a stable and regulated voltage, which is essential to guarantee reliable operation of both the sensors and the data acquisition system over time. The entire circuit is summarized in a block diagram, as shown in Figure 5.

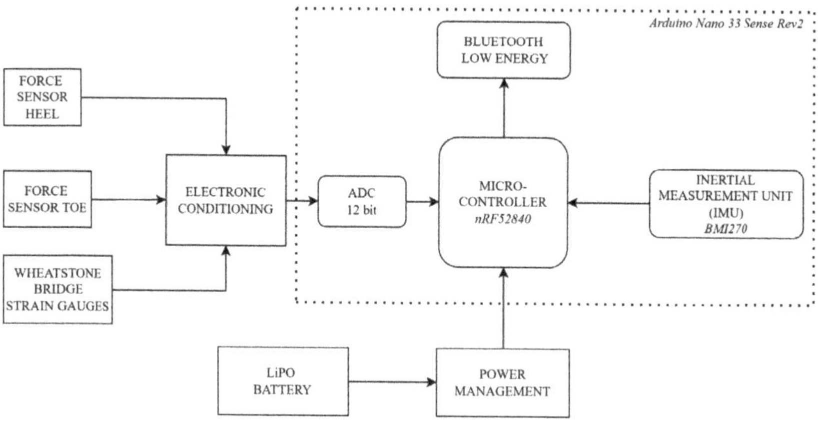

Figure 5. Block diagram of Hardware System.

The system is mounted as illustrated in Figure 6a. The PCB is enclosed in a 3D-printed case, as shown in Figure 6b, providing mechanical protection and integration of the battery and connectors. The enclosure also includes a power switch to turn the system on and off, as well as an LED indicator that lights up when the system is powered.

https://doi.org/10.3390/instruments10020023

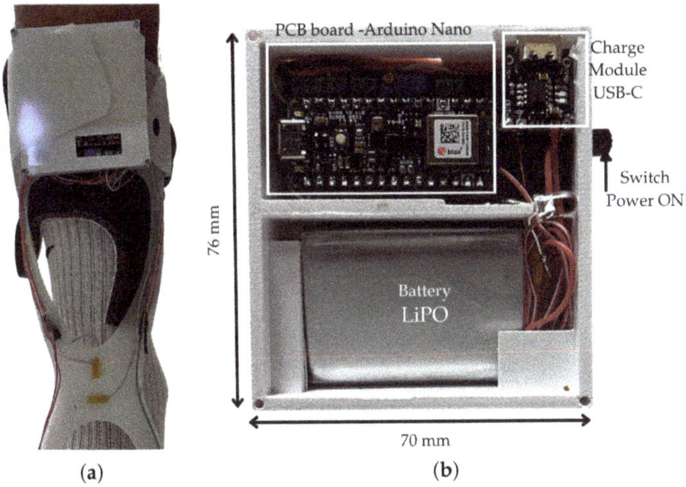

Figure 6. (**a**) Instrumented PLS-AFO developed, (**b**) Hardware System.

2.4. Microcontroller Firmware

The firmware logic implemented on the Arduino is shown in Figure 7. When the PCB board is powered, the Arduino initializes all variables and sets up the BLE characteristics, including the packet size and the variables to be sent. Each packet contains the following data:

- Timestamp (`uint32_t` base timestamp and `uint16_t` sample offsets; represents the acquisition time of each sample, ensuring accurate timing regardless of BLE transmission delays or PC clock differences);
- Gyroscope data (`int16_t` for each axis: gx, gy, gz);
- Accelerometer data (`int16_t` for each axis: ax, ay, az);
- Strain sensor output (`uint16_t`, 12-bit ADC value; represents the analog strain measurement and is converted to voltage in the Python-based GUI);
- Heel force sensor (`1-bit` ON/OFF; indicates contact with the ground);
- Toe force sensor (`1-bit` ON/OFF; indicates contact with the ground).

The firmware samples all sensors every 15 ms (66 Hz) and transmits data every 90 ms (10 Hz). The selection of these parameters is based on several considerations:

1. **BLE transmission intervals and communication range**: The 90 ms transmission interval represents a trade-off between real-time visualization and reliable BLE communication. Experimentally, higher transmission frequencies allow stable communication only up to about 1 m, whereas with the 90 ms interval, stable transmission is achieved up to 5–6 m. Therefore, the chosen interval ensures reliable data transmission and smooth visualization even at greater distances.
2. **Compatibility with human gait dynamics**: The 15 ms sampling interval (66 Hz) ensures sufficient temporal resolution according to the Nyquist criterion for signals related to human gait. This interval is sufficient to capture the main temporal features of the gait cycle, ensuring accurate representation of motion during walking [40].

When a device connects to the Arduino, a timer starts. Every 15 ms, the firmware reads the IMU, strain sensor, and FSR values and stores them in a buffer. The buffer can hold up to six packets, which takes approximately 90 ms to fill. When the buffer is full, the entire buffer is transmitted via BLE, and the buffer is reset.

https://doi.org/10.3390/instruments10020023

This approach ensures an effective sampling frequency of about 66 Hz, sufficient to capture human gait characteristics, while minimizing packet loss and enabling smooth real-time visualization in the Python-based GUI. The parameter selection reflects an experimentally verified trade-off between sampling frequency, BLE transmission interval, and communication range.

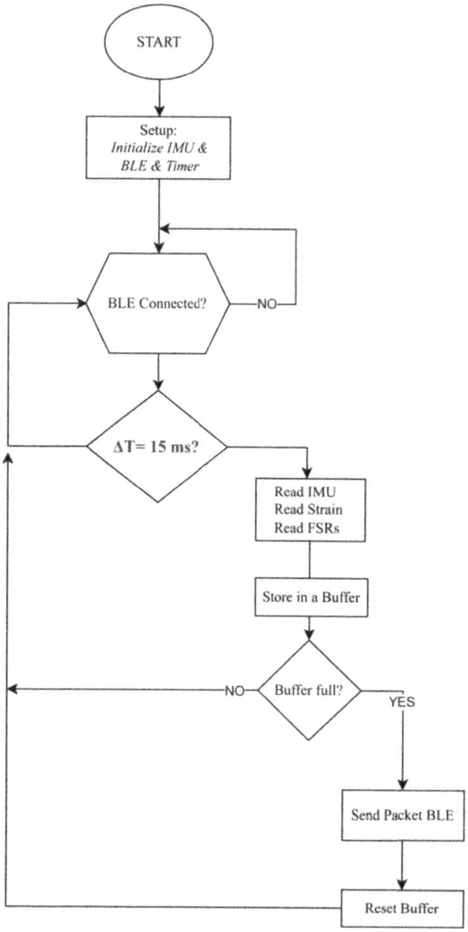

Figure 7. Arduino Firmware—Flowchart.

2.5. *Python-Based GUI*

The Python-based monitoring GUI (Figure 8) was developed using the PyQt6 framework to provide a real-time interface for the gait sensor node. The software architecture employs a multi-threaded approach to decouple the communication protocol management from the graphical rendering, ensuring a fluid data stream and responsive user interaction without lag.

BLE Communication and Buffered Reception. Wireless data acquisition is handled through the `bleak` library, operating in asynchronous mode to manage incoming Bluetooth notifications. To prevent processing bottlenecks, the application utilizes a dedicated background thread for continuous data reception and buffering. Raw binary data are extracted from 174-byte packets and transferred to the main GUI thread via a thread-safe queue.

https://doi.org/10.3390/instruments10020023

This architecture ensures that no samples are lost during high-frequency graphical updates while maintaining a stable connection.

Data Processing. A key design choice was to perform all signal scaling and physical unit conversions on the host PC rather than on the microcontroller. By transmitting raw integer ADC counts instead of floating-point values, the system significantly reduces the required BLE bandwidth and minimizes the computational overhead on the Arduino side. Each received packet consists of a hardware-generated reference timestamp followed by a burst of ten multi-channel samples. These include accelerometer and gyroscope readings (scaled into mg and dps), strain gauge data (converted into voltage), and the binary state of the Force Sensing Resistors (FSRs). The tilt angle (pitch) is then computed trigonometrically from the acceleration components. The use of hardware-side timestamps allows the system to reconstruct the time base with millisecond precision, eliminating timing errors caused by wireless transmission jitter.

Real-Time Visualization and Interface. The GUI simultaneously displays six synchronized plots arranged in a 2×3 grid, refreshed at 20 Hz. Three time-series panels show the dynamic trends of the inertial signals and the strain bridge voltage, while the remaining panels display the contact status of the force sensors, the pitch angle over time, and an analog gauge for immediate postural feedback. To optimize memory usage and CPU load, the software renders a rolling time window limited to the last 10 s of acquisition.

Recording and System Calibration. The application allows the entire acquisition session to be exported into CSV format for offline analysis. During the initial setup, a dedicated calibration mode provides real-time visual feedback of the strain gauge baseline voltage. This feature is critical for the operator to verify the Wheatstone bridge balance and adjust hardware gain stages before the start of dynamic gait trials.

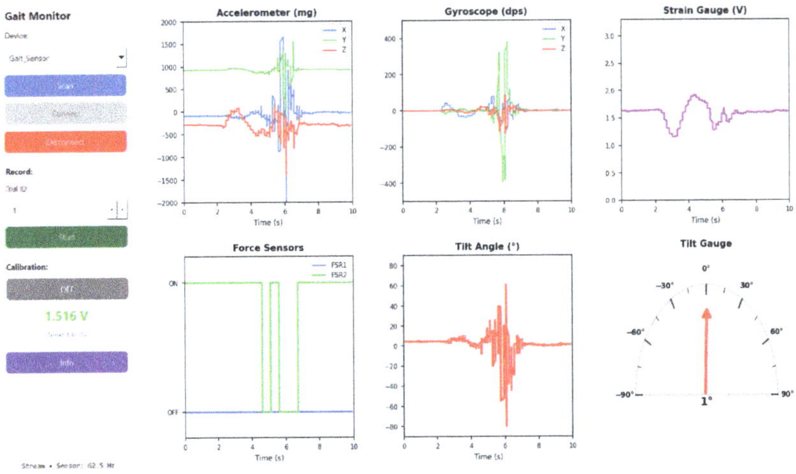

Figure 8. Python-based GUI.

2.6. Validation System—Optical System

System validation was conducted using a gold-standard optical three-dimensional motion capture system. The reference system was the Smart-DX EVO (BTS Bioengineering), composed of ten infrared cameras covering the entire acquisition volume and two force platforms for ground reaction force measurements.

The system is marker-based and employs passive reflective markers placed on anatomical landmarks. Each camera emits infrared light that is reflected by the markers and

https://doi.org/10.3390/instruments10020023

detected by multiple cameras. Through spatial triangulation, the three-dimensional coordinates of each marker are reconstructed with high precision.

Optical motion capture systems are considered the gold standard in biomechanical analysis due to their high spatial resolution, elevated sampling frequency, and validated accuracy in three-dimensional kinematic reconstruction. These systems are extensively used in clinical gait analysis and research settings, providing reliable and repeatable measurements of joint kinematics and kinetics [41].

Marker placement was performed according to the Rizzoli protocol [42], as illustrated in Figure 9. In configuration (a), reflective markers were positioned exclusively on the orthosis to quantify plantarflexion–dorsiflexion angular displacement and to calibrate and validate the strain gauge measurements against the optical motion capture system. In configuration (b), three markers were placed on the knee according to the same protocol to define a local coordinate system. The inertial measurement unit (IMU) was validated against the optical motion capture system by comparing the measured tilt angle associated with postural imbalance and fall risk.

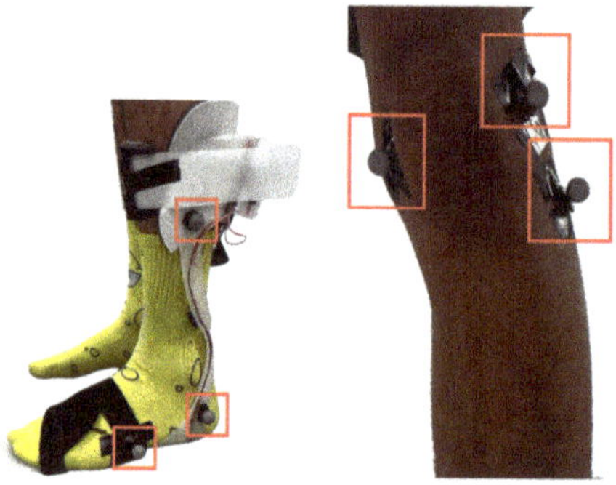

(a) (b)

Figure 9. (a) Marker measuring Plantar-Dorsi Flexion, (b) Marker measuring Tilt Angle.

3. Results

3.1. FSR Validation

Initial tests were performed on the force-sensitive resistors (FSRs) using a dynamometer to determine appropriate threshold values for detecting foot-ground contact. Table 1 shows representative resistance measurements for selected forces, covering the range relevant for walking.

Table 1. FSR resistance vs. applied force (heel sensor).

Force (N)	Resistance (kΩ)
0	9.74
41.3	5.3
88.8	3.72
115	3.13
204	1.96
322	1.01
491	0.782

Based on these measurements, resistance thresholds of 5 kΩ for both the heel and toe sensors were selected to reliably indicate foot-ground contact. In the data acquisition system, these values correspond approximately to 3.3 V.

https://doi.org/10.3390/instruments10020023

The validation of the FSR-based step detection is presented in Figure 10 and Table 2. Figure 10 shows the comparison between foot-ground contact detected by the force platform and the FSR signal. Ground contact from the force platform is identified using a threshold of approximately 40 N to filter out noise and ensure reliable detection [38,39], while the FSR signal is considered active whenever at least one sensor (heel or toe) exceeds its resistance threshold. Vertical dashed lines indicate the heel strike and toe-off events detected by the force platform.

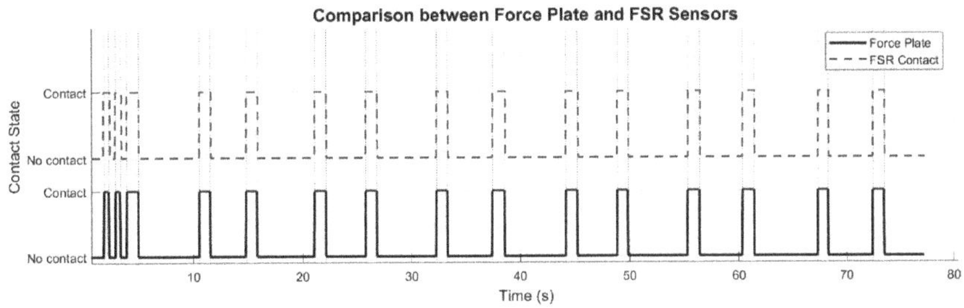

Figure 10. Comparison between foot-ground contact detected by the force platform and the FSR sensors. Black line: force platform; blue line: OR logic of heel and toe FSRs. Vertical dashed lines indicate heel strike and toe off events.

Table 2 reports the mean delays between the FSR detections and the force platform events. The heel sensor detected heel strike slightly earlier than the platform, with a mean delay of -0.031 ± 0.046 s, whereas the toe sensor detected toe off slightly later, with a mean delay of 0.012 ± 0.056 s.

To better understand the timing differences in relation to the stance phase, the delays were expressed as a percentage of the stance duration using the following equation:

$$\text{Normalized delay (\%)} = \frac{\text{Delay (s)}}{\text{Stance duration (s)}} \times 100 \quad (1)$$

The observed anticipations or delays (2.8% for heel strike and 1.1% for toe off, see Table 2) indicate that the FSR sensors detect gait events with minimal temporal deviation relative to the stance phase. These differences may be influenced by factors such as foot size, orthosis dimensions, and slight variations in sensor placement, which can cause a forward or backward shift in event detection. Overall, the FSR sensors demonstrated excellent temporal agreement with the force platform, confirming the validity of the selected sensor locations.

Table 2. FSR validation results with normalized delays.

FSR Position	Delay (s)	Std Deviation (s)	Normalized Delay (%)
Heel	−0.031	0.046	2.8
Toe	0.012	0.056	1.1

3.2. Strain Gauge Positioning

As described in the previous section, two configurations were tested: a half Wheatstone bridge and a full Wheatstone bridge. In both cases, two movements were performed: a lunge to measure dorsiflexion and the opposite movement to measure plantarflexion, as shown in Figure 11. At the beginning, subjects were asked to hold a steady position for 10 s to allow measurement and to synchronize the two systems at a 0° angle.

https://doi.org/10.3390/instruments10020023

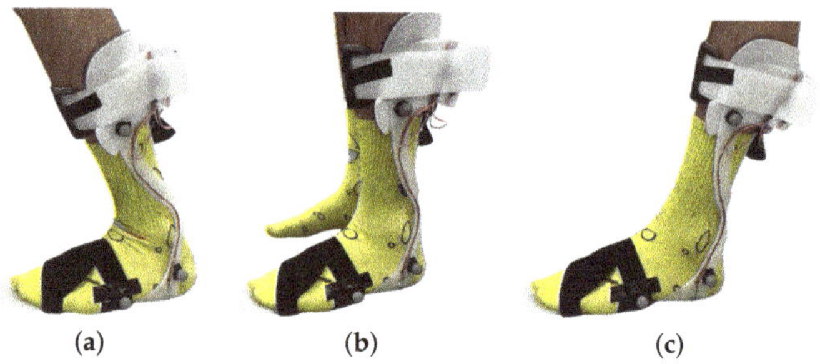

Figure 11. (**a**) Dorsiflexion, (**b**) Steady, (**c**) Plantarflexion.

Figure 12 shows the behavior of the configuration with four strain gauges. In this case, when the deformation angle decreases beyond a certain value, the output signal does not follow the expected trend. Instead of decreasing, the signal begins to increase.

This behavior is mainly related to the orientation of the strain gauges and to the mechanical response of the orthosis structure. During dorsiflexion, under ideal conditions, the gauges should exhibit predictable strain patterns according to beam theory: on the outer side of the orthosis, the longitudinal gauge experiences tensile strain (positive ε), while the transversal gauge undergoes compressive strain due to the Poisson effect. Conversely, on the inner side, the longitudinal gauge is compressed while the transversal gauge is stretched.

However, beyond a certain angular threshold, both transversal gauges exhibit a strain reversal. This phenomenon is not attributable to the material properties but rather to the curved geometry of the orthosis structure. The orthosis is made of polypropylene (PP), a thermoplastic polymer with a Young's modulus of approximately 0.8 GPa and a Poisson's ratio of 0.41. While these material properties influence the deformation behavior, the primary cause of the strain reversal lies in the three-dimensional deformation of the curved structure. As the bending angle increases, the curved geometry induces complex stress distributions that cause the transversal strain components to reverse direction on both the outer and inner surfaces, deviating from the simple Poisson effect predicted by linear beam theory.

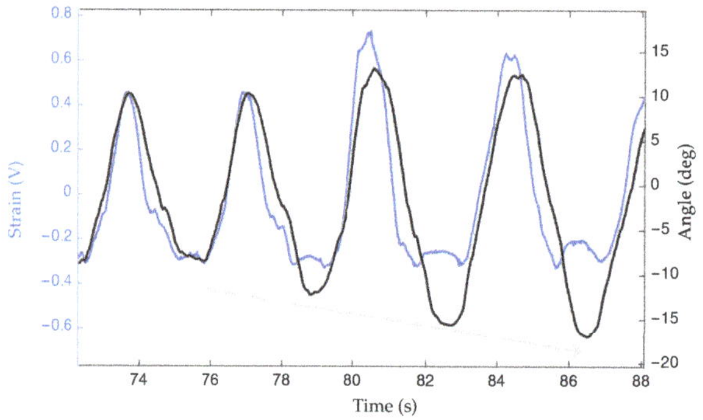

Figure 12. Angle vs. strain during a squat movement using a full Wheatstone bridge. The measured value is shown in black, while the reference system is also indicated.

After this test, only the two strain gauges oriented in the longitudinal direction were used in a half Wheatstone bridge configuration to measure the deformation of the orthosis.

https://doi.org/10.3390/instruments10020023

3.3. Strain Gauge Calibration

A calibration procedure was conducted to establish a linear relationship between the strain gauge output voltage and the ankle joint angle measured by the optoelectronic system. A total of sixteen measurement sessions were performed across three subjects. Each session included ten plantarflexion and ten dorsiflexion movements, as shown in Figure 13. This protocol enabled the evaluation of both intra-subject and inter-subject variability, as well as the repeatability and robustness of the system under realistic conditions.

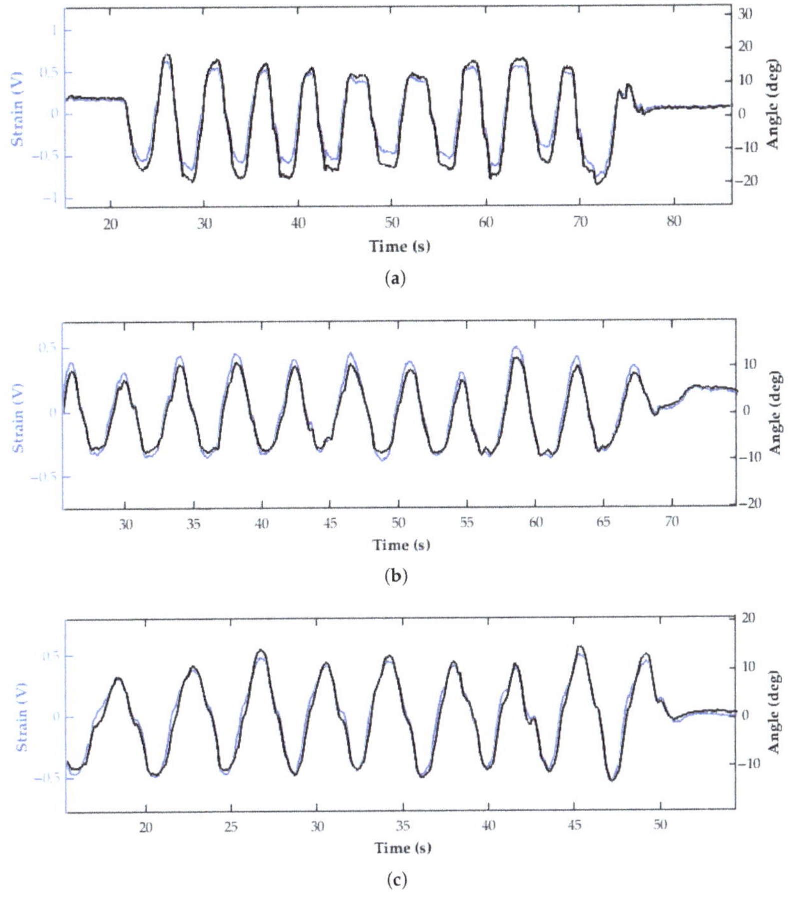

Figure 13. Movements of the three subjects: (**a**) Subject 1, (**b**) Subject 2, (**c**) Subject 3. The system output is shown in blue, while the reference signal acquired from the BTS system is shown in black.

For each subject, the calibration data were processed independently. The strain signal and the optoelectronic reference angle were first temporally synchronized by optimizing the time shift to maximize the coefficient of determination. Both signals were then mean-centered to remove static offsets, and a zero-intercept linear regression was performed considering only data within the angular range of $\pm 15^{\circ}$.

The results, shown in Figure 14, demonstrate consistent linearity across all subjects. The average sensitivity was approximately 33 mV/deg, with a pooled coefficient of determination (R^2) of 98.43% and an RMSE of 36.33 mV, corresponding to an angular error of approximately 1°. These results confirm that the half Wheatstone bridge configuration with

https://doi.org/10.3390/instruments10020023

longitudinal strain gauges provides a linear and repeatable response suitable for real-time monitoring of orthosis deformation during functional movements.

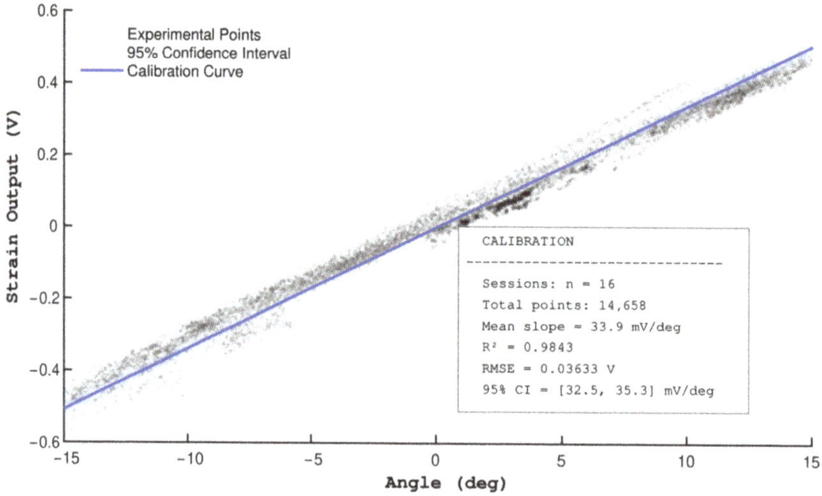

Figure 14. Calibration Curve BTS Angle vs. Strain.

3.4. Strain Gauge Validation

The validation of the previously calibrated system was conducted on ten subjects. The same experimental setup of the optoelectronic system was maintained for all trials. Each subject performed walking trials within the acquisition volume of the optical motion capture system, as illustrated in Figure 15 for four representative subjects.

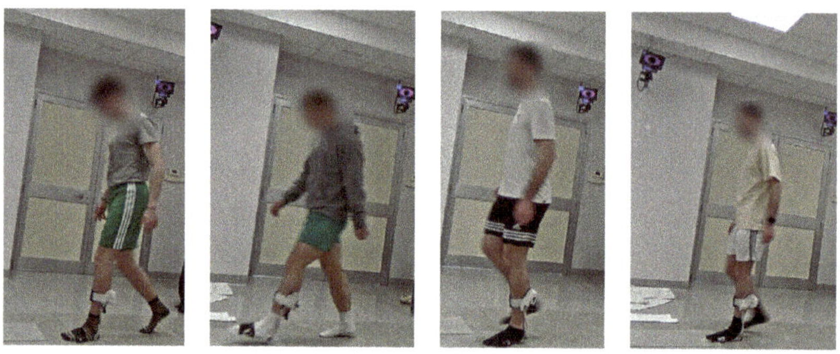

Figure 15. Walking trials.

Figure 16 presents two representative gait cycles for four selected subjects, illustrating the system's performance across different accuracy levels. The discrepancies observed between the strain gauge measurements and the reference system (see Table 3) can be attributed to several factors that may contribute to an overestimation of the actual measurement error.

During static calibration trials, the optoelectronic system maintained continuous marker visibility and provided highly reliable reference measurements; however, during dynamic gait trials, certain factors introduced additional measurement uncertainty, as the system became susceptible to occasional measurement artifacts such as marker flickering or temporary loss of signal due to occlusions caused by the orthosis structure or the subject's limbs masking the markers from the cameras [43,44], and, additionally, one marker is

https://doi.org/10.3390/instruments10020023

positioned directly on the foot, making it particularly prone to artifacts due to impacts with the ground and rapid movements during gait, which can further affect the accuracy and stability of the recorded measurements.

Furthermore, the different sampling rates of the two systems—66 Hz for strain gauge acquisition versus 100 Hz for the optoelectronic system—necessitate resampling and temporal alignment for correlation. This process can introduce synchronization errors, particularly during high-velocity phases of the gait cycle, adding a systematic source of uncertainty to the comparison.

Given these considerations, the reported mean RMSE of 1.65° likely represents a conservative estimate that includes both the intrinsic error of the strain gauge system and measurement artifacts from the dynamic validation protocol. Even with this conservative estimate, the performance is significantly superior to that of alternative systems used to measure ankle–foot angle, such as IMU-based approaches, where RMSE values typically exceed 5° [45–47].

Beyond accuracy, the proposed system offers substantial advantages in user comfort and setup efficiency. Conventional IMU-based approaches require at least two sensors—one on the foot and one on the shank—to estimate dorsiflexion and plantarflexion angles, introducing risks of sensor misalignment and skin movement artifacts. In contrast, the proposed solution relies on only two strain gauges mounted directly on the orthosis, thereby reducing hardware complexity, eliminating sensor-skin interface issues, and simplifying the overall measurement setup while maintaining superior accuracy.

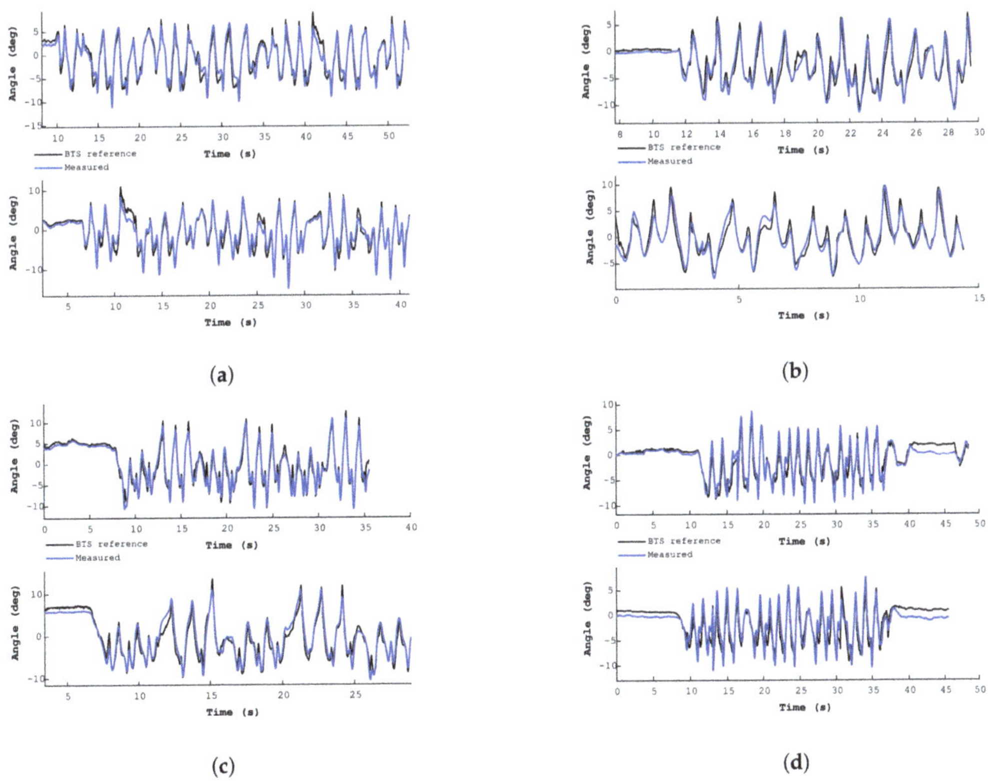

Figure 16. Gait Validation Results for Four Subjects. (**a**) Subject 1, (**b**) Subject 2, (**c**) Subject 3, (**d**) Subject 4.

https://doi.org/10.3390/instruments10020023

Table 3. Gait validation results per subject. Mean values across all subjects are 1.65 ± 0.32° for RMSE and 0.85 ± 0.06 for R^2 (mean ± SD).

Subject	Shoe Size	RMSE (°)	SD (°)	R^2
Subject 1	42	1.506	0.223	0.8951
Subject 2	41	1.580	0.236	0.7760
Subject 3	43	1.381	0.225	0.9112
Subject 4	42	2.077	0.297	0.8600
Subject 5	46	1.464	0.250	0.8062
Subject 6	46	1.290	0.143	0.8657
Subject 7	44	1.935	0.278	0.8454
Subject 8	43	1.564	0.105	0.9167
Subject 9	46	2.239	0.143	0.7180
Subject 10	43	1.473	0.160	0.8621

3.5. IMU—Validation

IMUs are widely investigated and characterized in wearable sensing applications for human motion analysis [48–53]. Therefore, this study does not aim to validate the IMU technology itself but rather to assess the correctness of its positioning and implementation within the proposed setup.

A controlled static tilt movement was performed, and the angle estimated by the IMU mounted on the Arduino was compared with the reference standard. The comparison (Figure 17) shows a discrepancy between the two systems, with an RMSE of 5.88 ± 1.33 degrees.

This difference is mainly due to the indirect nature of the IMU measurement. The tilt angle is estimated through sensor fusion of accelerometer and gyroscope signals, which are affected by noise, bias, drift, and possible misalignment with respect to the anatomical reference frame. Despite this, the observed error is consistent with values commonly reported in the literature for wearable IMU-based systems [49].

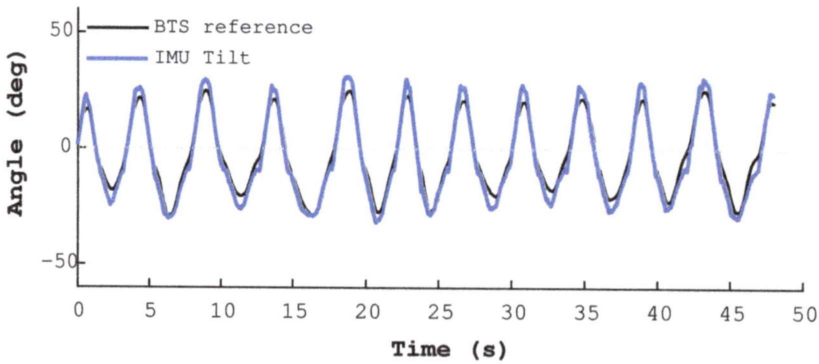

Figure 17. Validation BTS Angle vs. IMU tilt.

3.6. Gait Cycle Analysis

This section explains how the developed system can be used in a clinical scenario, with particular focus on how gait parameters can be extracted from the acquired signals.

Figure 18 shows several gait cycles performed by one subject. The force sensors are used to detect each gait cycle and to identify the main gait events. At the same time, the strain gauges measure the orthosis deformation during walking.

In this study, a positive ankle angle corresponds to plantarflexion, while a negative angle corresponds to dorsiflexion. To correctly segment the gait cycle, 0% is defined at the

https://doi.org/10.3390/instruments10020023

heel strike event, and the cycle ends at the subsequent heel strike of the same foot. After the initial heel strike (0%), toe contact occurs at approximately 14% of the gait cycle. Heel-off occurs around 43%, followed by toe-off at approximately 52%. These events allow a clear identification of the stance and swing phases.

In addition to the ankle angle, the IMU sensor provides a measure of the Lateral Shank Tilt Angle, which describes how much the leg inclines mediolaterally during the gait cycle. A positive value indicates a lateral inclination, while a negative value indicates a medial inclination. As shown in Figure 18b, the shank exhibits a characteristic oscillation between lateral and medial tilt throughout the gait cycle, with peak lateral tilt occurring during the swing phase and peak medial tilt during mid-stance.

By combining the signals obtained from the two strain gauges, the two force sensors, and the IMU, several clinically relevant parameters can be extracted. From the force sensors, temporal parameters such as stance duration, swing duration, and timing of gait events can be determined. From the orthosis deformation, kinematic parameters including peak plantarflexion, peak dorsiflexion, and ankle range of motion can be estimated. From the IMU, the mediolateral shank inclination profile across the gait cycle can be monitored.

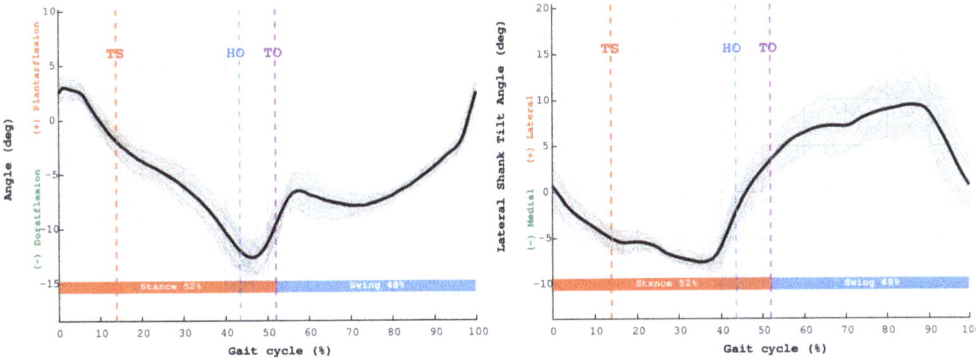

(a) (b)

Figure 18. Gait cycle showing the main events—TS (Toe Strike), HO (Heel Off), and TO (Toe Off) (**a**) Ankle plantarflexion–dorsiflexion profile. (**b**) Lateral Shank Tilt Angle during the gait cycle.

Table 4 summarizes the parameters that can be monitored using only two strain gauges, two force sensors, and one IMU. These results demonstrate that the proposed system can provide meaningful clinical information with a limited number of sensors, making it suitable for practical and low-cost clinical applications.

Table 4. Typical gait cycle parameters (mean ± SD). Ankle angle: positive = plantarflexion, negative = dorsiflexion. Lateral Shank Tilt Angle (left shank): positive = lateral, negative = medial.

Gait Event/Parameter	Mean/Position	SD
Heel Strike (HS)	0%	–
Toe Strike (TS)	14.1%	±2.9%
Heel Off (HO)	43.7%	±4.3%
Toe Off (TO)	52.1%	±2.4%
Heel Strike (HS)	100%	–
Cycle Duration	1.57 s	±0.10 s
Cadence	76.4 steps/min	–

https://doi.org/10.3390/instruments10020023

Table 4. *Cont.*

Gait Event/Parameter	Mean/Position	SD
Ankle Angle		
Peak Plantarflexion	+3.0 deg @ 1%	±1.1 deg
Peak Dorsiflexion	−12.6 deg @ 46%	±1.3 deg
Lateral Shank Tilt Angle		
Peak Lateral	+9.5 deg @ 86%	±3.5 deg
Peak Medial	−7.6 deg @ 35%	±0.9 deg

4. Conclusions

This study designed, realized, characterized, and validated an instrumented ankle–foot orthosis for gait monitoring, integrating deformation sensing, inertial measurements, and ground-contact detection. The system was developed with the dual goal of providing clinically relevant biomechanical information and maintaining practical usability in real-world rehabilitation settings.

The orthosis deformation was measured using two longitudinal strain gauges to estimate ankle plantar- and dorsiflexion, with validation against a gold-standard optoelectronic motion capture system. The strain-gauge-based approach achieved an RMSE of approximately 1.6 degrees, outperforming the IMU in angle estimation while reducing circuit complexity and maintaining high measurement accuracy. These results confirm that embedding strain gauges directly into the orthosis structure is an effective strategy for continuous joint angle monitoring during locomotion.

Two force sensors were positioned on the orthosis to detect ground contact and to identify the stance and swing phases of the gait cycle. Validation against a force platform confirmed the correct placement of the sensors and their reliability in detecting contact and non-contact events. Accurate gait phase segmentation is a key requirement for a wide range of clinical analyses, including step detection and cadence estimation in patients with pathological conditions.

The IMU was used to measure lateral shank tilt, providing complementary information for safety monitoring and the assessment of balance-related deviations. Large deviations in lateral tilt may indicate instability or an increased risk of falling, highlighting the clinical value of combining multiple sensing modalities.

The instrumented AFO communicates via Bluetooth Low Energy (BLE) with a custom Python-based graphical interface for real-time data acquisition, visualization, and storage. The wireless architecture removes cable-related constraints, preserving natural gait mechanics and improving patient comfort during prolonged monitoring sessions.

The electronic architecture and data acquisition capabilities of the proposed system were evaluated using healthy participants, demonstrating its technical feasibility for gait monitoring. Future studies on clinical populations will be required to extend validation toward rehabilitation assessment and orthosis-related applications.

Author Contributions: Conceptualization, M.S.; methodology, M.S. and S.M.; software, S.M.; validation, S.M.; formal analysis, S.M.; investigation, S.M.; resources, S.M.; data curation, S.M.; writing—original draft preparation, S.M.; writing—review and editing, S.M.; supervision, M.S.; project administration, M.S.; funding acquisition, M.S. All authors have read and agreed to the published version of the manuscript.

Funding: This work has received funding from the European Union's Horizon Europe research and innovation programme under Grant Agreement No. 101136376 (SmILE—Smart Implants for Life Enrichment, HORIZON-HLTH-2023-STAYHLTH-01, CUP: D83C24001650006). The views and opinions expressed are, however, those of the authors only and do not necessarily reflect those of the European Union or the European Health and Digital Executive Agency (HaDEA). Neither the European Union nor the granting authority can be held responsible for them.

Informed Consent Statement: The University Research Ethics Committee (CERA) of the University of Brescia approved this testing procedure. Informed consent was obtained from all subjects involved in the study.

Data Availability Statement: The data supporting the findings of this study are available from the corresponding author upon reasonable request. Requests should be made using the contact information provided in this article.

Conflicts of Interest: The authors declare no conflicts of interest.

References

1. Kharb, A.; Saini, V.; Jain, Y.K.; Dhiman, S. A review of gait cycle and its parameters. *Int. J. Comput. Eng. Manag.* **2011**, *13*, 78–83.
2. Klaewkasikum, K.; Patathong, T.; Angsanuntsukh, C.; Woratanarat, T.; Sanguantrakul, J.; Woratanarat, P. The ankle kinematic reference of normal gait pattern in Thai adults. *Front. Surg.* **2022**, *9*, 915090. [CrossRef] [PubMed]
3. Shefa, F.R.; Sifat, F.H.; Shah, S.C.; Kibria, M.G. IoT-Based Smart Ankle-Foot Orthosis for Patients with Gait Imbalance. In *Proceedings of the 2023 23rd International Conference on Control, Automation and Systems (ICCAS)*; IEEE: New York, NY, USA, 2023; pp. 969–974. [CrossRef]
4. Subramaniam, S.; Faisal, A.I.; Deen, M.J. Wearable Sensor Systems for Fall Risk Assessment: A Review. *Front. Digit. Health* **2022**, *4*, 921506. [CrossRef]
5. Resch, S.; Kousha, A.; Carroll, A.; Severinghaus, N.; Rehberg, F.; Zatschker, M.; Söyleyici, Y.; Sanchez-Morillo, D. Smart Device Development for Gait Monitoring: Multimodal Feedback in an Interactive Foot Orthosis, Walking Aid, and Mobile Application. *Technologies* **2025**, *13*, 588. [CrossRef]
6. Sekiyama, M.; Homma, K.; Wakita, Y.; Kajitani, I.; Mori, I.; Matsumoto, Y. Investigation of IoT on ankle-foot orthosis by introducing an existing life-activity measurement system. In *Proceedings of the 2021 IEEE International Conference on Advanced Robotics and Its Social Impacts (ARSO)*; IEEE: New York, NY, USA, 2021; pp. 246–253. [CrossRef]
7. Abtahi, M.; Gyllinsky, J.V.; Paesang, B.; Barlow, S.; Constant, M.; Gomes, N.; Tully, O.; D'Andrea, S.E.; Mankodiya, K. MagicSox: An E-textile IoT system to quantify gait abnormalities. *Smart Health* **2018**, *5–6*, 4–14. [CrossRef]
8. Sardini, E.; Serpelloni, M.; Lancini, M.; Pasinetti, S. Wireless Instrumented Crutches for Force and Tilt Monitoring in Lower Limb Rehabilitation. *Procedia Eng.* **2014**, *87*, 348–351. [CrossRef]
9. Rathke, C.L.; Pimentel, V.C.d.A.; Alsina, P.J.; do Espírito Santo, C.C.; Dantas, A.F.O.d.A. IoT-Based Wireless System for Gait Kinetics Monitoring in Multi-Device Therapeutic Interventions. *Sensors* **2024**, *24*, 5799. [CrossRef] [PubMed]
10. Feng, Y.; Liu, Y.; Fang, Y.; Chang, J.; Deng, F.; Liu, J.; Xiong, Y. Advances in the application of wearable sensors for gait analysis after total knee arthroplasty: A systematic review. *Arthroplasty* **2023**, *5*, 49. [CrossRef]
11. Porciuncula, F.; Roto, A.V.; Kumar, D.; Davis, I.; Roy, S.; Walsh, C.J.; Awad, L.N. Wearable Movement Sensors for Rehabilitation: A Focused Review of Technological and Clinical Advances. *PM R J. Inj. Funct. Rehabil.* **2018**, *10*, S220–S232. [CrossRef]
12. Gu, B.; Kim, H.S.; Kim, H.; Yoo, J.I. Advancements in Wearable Sensor Technologies for Health Monitoring in Terms of Clinical Applications, Rehabilitation, and Disease Risk Assessment: Systematic Review. *JMIR Mhealth Uhealth* **2026**, *14*, e76084. [CrossRef]
13. Zarezade, R.; Arazpour, M.; Aminian, G. The effect of anterior ankle-foot orthosis and posterior ankle-foot orthosis on functional ambulation in stroke patients. *J. Rehabil. Assist. Technol. Eng.* **2022**, *9*, 205566832210824. [CrossRef]
14. Ganokratanaa, T.; Pramkeaw, P.; Ketcham, M.; Khamphao, Y.; Wanthanasak, P. Intelligent IoT-enabled Ankle Brace: Enhancing Mobility and Safety. In *Proceedings of the 2024 IEEE International Conference on Cybernetics and Innovations (ICCI)*; IEEE: New York, NY, USA, 2024; pp. 1–12. [CrossRef]
15. Sardini, E.; Serpelloni, M.; Lancini, M. Wireless Instrumented Crutches for Force and Movement Measurements for Gait Monitoring. *IEEE Trans. Instrum. Meas.* **2015**, *64*, 3369–3379. [CrossRef]
16. Sloss, R. The effects of foot orthoses on the ground reaction forces during walking. Part 1. *Foot* **2001**, *11*, 205–214. [CrossRef]
17. Svensson, W.; Salomonsson, T.; Holmberg, U. Foot orthosis strain sensing in hill walking. In *Proceedings of the 2007 IEEE 10th International Conference on Rehabilitation Robotics*; IEEE: New York, NY, USA, 2007; pp. 700–703. [CrossRef]
18. Cantù, E.; Fapanni, T.; Giorgi, G.; Narduzzi, C.; Sardini, E.; Serpelloni, M.; Tonello, S. Printed Multi-EMG Electrodes on the 3D Surface of an Orthosis for Rehabilitation: A Feasibility Study. *IEEE Sens. J.* **2021**, *21*, 14407–14417. [CrossRef]
19. Hamid, A.; Patar, M.N.A.A.; Ayub, M.A. Force Sensor Detection and Performance Evaluation of New Active System Ankle Foot Orthosis. *Procedia Eng.* **2012**, *41*, 510–515. [CrossRef]

https://doi.org/10.3390/instruments10020023

20. Gil-Castillo, J.; Alnajjar, F.; Koutsou, A.; Torricelli, D.; Moreno, J.C. Advances in neuroprosthetic management of foot drop: A review. *J. Neuroeng. Rehabil.* **2020**, *17*, 46. [CrossRef]
21. Rescio, G.; Leone, A.; Francioso, L.; Siciliano, P. Sensorized Insole for Diabetic Foot Monitoring. *Proceedings* **2018**, *2*, 860. [CrossRef]
22. Tanino, G.; Tomita, Y.; Mizuno, S.; Maeda, H.; Miyasaka, H.; Orand, A.; Takeda, K.; Sonoda, S. Development of an ankle torque measurement device for measuring ankle torque during walking. *J. Phys. Ther. Sci.* **2015**, *27*, 1477–1480. [CrossRef]
23. Lin, J.; Yu, J.; Chen, F.; Jia, X.; Li, J.; Zhao, J.; He, J.; Hou, X.; Mu, J.; Chou, X. A high-performance flexible coaxial BTO/PAN piezoelectric nanofibers for facial expression analysis. *Chem. Eng. J.* **2025**, *516*, 163932. [CrossRef]
24. Hou, X.; Zhang, S.; Yu, J.; Cui, M.; He, J.; Li, L.; Wang, X.; Chou, X. Flexible Piezoelectric Nanofibers/Polydimethylsiloxane-Based Pressure Sensor for Self-Powered Human Motion Monitoring. *Energy Technol.* **2020**, *8*, 1901242. [CrossRef]
25. Li, Y.; Song, H.; Cao, J.; Qiu, X.; Xu, Q.; Yang, Z.; Cheng, J. A pre-strained WPU/Gr/Ecoflex composite-based screen-printed flexible resistive strain sensor for gesture motion recognition. *Sens. Actuators A Phys.* **2025**, *396*, 117214. [CrossRef]
26. Yu, J.; Xian, S.; Mu, J.; Wang, M.; Wang, Y.; Hou, X.; Zhang, L.; He, J.; Mu, J.; Chou, X. Hybrid electromechanical properties of hetero-doped and homogeneously bonded dual-mode pressure sensor for indoor body area network node. *Sci. China Inf. Sci.* **2023**, *67*, 112401. [CrossRef]
27. Borghetti, M.; Serpelloni, M.; Sardini, E. Printed Strain Gauge on 3D and Low-Melting Point Plastic Surface by Aerosol Jet Printing and Photonic Curing. *Sensors* **2019**, *19*, 4220. [CrossRef]
28. Bellitti, P.; Borghetti, M.; Cantù, E.; Sardini, E.; Serpelloni, M. Resistive Sensors for Smart Objects: Analysis on Printing Techniques. *IEEE Trans. Instrum. Meas.* **2022**, *71*, 9507715. [CrossRef]
29. Aminian, K.; Najafi, B.; Büla, C.; Leyvraz, P.F.; Robert, P. Spatio-temporal parameters of gait measured by an ambulatory system using miniature gyroscopes. *J. Biomech.* **2002**, *35*, 689–699. [CrossRef] [PubMed]
30. Betz, J.; Klingspor, C.; Seel, T. IMU-based Assessment of Ankle Inversion Kinematics and Orthosis Migration. In *Proceedings of the 2019 41st Annual International Conference of the IEEE Engineering in Medicine and Biology Society (EMBC)*; IEEE: New York, NY, USA, 2019; pp. 6395–6400. [CrossRef]
31. Seel, T.; Raisch, J.; Schauer, T. IMU-Based Joint Angle Measurement for Gait Analysis. *Sensors* **2014**, *14*, 6891–6909. [CrossRef]
32. Wang, F.; Liang, W.; Afzal, H.M.R.; Fan, A.; Li, W.; Dai, X.; Liu, S.; Hu, Y.; Li, Z.; Yang, P. Estimation of Lower Limb Joint Angles and Joint Moments during Different Locomotive Activities Using the Inertial Measurement Units and a Hybrid Deep Learning Model. *Sensors* **2023**, *23*, 9039. [CrossRef]
33. Lv, J.; Ravankar, A.A.; Kobayashi, Y.; Emaru, T. A method of low-cost IMU calibration and alignment. In *Proceedings of the 2016 IEEE/SICE International Symposium on System Integration (SII)*; IEEE: New York, NY, USA, 2016; pp. 373–378. [CrossRef]
34. Niswander, W.; Jara, R.; Bazan, J.; Kujala, S.; Ribeiro, P.; Gonzalez, R.; Perez, J.; Roa, M. Optimization of IMU Sensor Placement for the Measurement of Lower Limb Joint Kinematics. *Sensors* **2020**, *20*, 5993. [CrossRef]
35. Chong, Y.W.; Ismail, W.; Ko, K.; Lee, C.Y. Energy Harvesting For Wearable Devices: A Review. *IEEE Sens. J.* **2019**, *19*, 9047–9062. [CrossRef]
36. Huang, X.; Xue, Y.; Ren, S.; Wang, F. Sensor-Based Wearable Systems for Monitoring Human Motion and Posture: A Review. *Sensors* **2023**, *23*, 9047. [CrossRef] [PubMed]
37. Mahraoui, S.; Serpelloni, M. Wireless Instrumented Ankle Foot Orthosis (AFO) for Gait Cycle Monitoring: a Preliminary Study. In *Proceedings of the 2025 IEEE International Workshop on Metrology for Industry 4.0 & IoT (MetroInd4.0 & IoT)*; IEEE: New York, NY, USA, 2025; pp. 89–94. [CrossRef]
38. Zeni, J.; Richards, J.; Higginson, J. Two simple methods for determining gait events during treadmill and overground walking using kinematic data. *Gait Posture* **2008**, *27*, 710–714. [CrossRef] [PubMed]
39. Hreljac, A.; Marshall, R.N. Algorithms to determine event timing during normal walking using kinematic data. *J. Biomech.* **2000**, *33*, 783–786. [CrossRef] [PubMed]
40. Raffalt, P.C.; McCamley, J.; Denton, W.; Yentes, J.M. Sampling frequency influences sample entropy of kinematics during walking. *Med. Biol. Eng. Comput.* **2019**, *57*, 759–764. [CrossRef]
41. Piche, E.; Guilbot, M.; Chorin, F.; Guerin, O.; Zory, R.; Gerus, P. Validity and repeatability of a new inertial measurement unit system for gait analysis on kinematic parameters: Comparison with an optoelectronic system. *Measurement* **2022**, *198*, 111442. [CrossRef]
42. Hausselle, J.; Haddox, A.; Kasitz, J.; Azoug, A. An interventional exploratory study to assess the effect of footwear on postural stability and strategy during quiet standing. *Int. Biomech.* **2021**, *8*, 63–74. [CrossRef]
43. Benoit, D.L.; Ramsey, D.K.; Lamontagne, M.; Xu, L.; Wretenberg, P.; Renström, P. Effect of skin movement artifact on knee kinematics during gait and cutting motions measured in vivo. *Gait Posture* **2006**, *24*, 152–164. [CrossRef]
44. Moro, M.; Marchesi, G.; Hesse, F.; Odone, F.; Casadio, M. Markerless vs. Marker-Based Gait Analysis: A Proof of Concept Study. *Sensors* **2022**, *22*, 2011. [CrossRef]

https://doi.org/10.3390/instruments10020023

45. Bissenov, A.; Zhubi, E.; Engh, M.e.a. Concurrent validity of wearable IMUs for sagittal plane lower-limb range of motion during walking and estimated ground reaction forces: A systematic review and meta-analysis. *J. Orthop. Surg. Res.* **2025**, *20*, 891. [CrossRef] [PubMed]
46. Shuai, Z.; Dong, A.; Liu, H.; Cui, Y. Reliability and Validity of an Inertial Measurement System to Quantify Lower Extremity Joint Angle in Functional Movements. *Sensors* **2022**, *22*, 863. [CrossRef]
47. Mundt, M.; Koeppe, A.; David, S.; Witter, T.; Bamer, F.; Potthast, W.; Markert, B. Estimation of Gait Mechanics Based on Simulated and Measured IMU Data Using an Artificial Neural Network. *Front. Bioeng. Biotechnol.* **2020**, *8*, 41. [CrossRef]
48. McGrath, T.; Stirling, L. Body-Worn IMU-Based Human Hip and Knee Kinematics Estimation during Treadmill Walking. *Sensors* **2022**, *22*, 2544. [CrossRef] [PubMed]
49. Leardini, A.; Chiari, L.; Croce, U.D.; Cappozzo, A. Human movement analysis using stereophotogrammetry: Part 3. Soft tissue artifact assessment and compensation. *Gait Posture* **2005**, *21*, 212–225. [CrossRef] [PubMed]
50. Cerfoglio, S.; Capodaglio, P.; Rossi, P.; Conforti, I.; D'Angeli, V.; Milani, E.; Galli, M.; Cimolin, V. Evaluation of Upper Body and Lower Limbs Kinematics through an IMU-Based Medical System: A Comparative Study with the Optoelectronic System. *Sensors* **2023**, *23*, 6156. [CrossRef]
51. Mao, Y.; Ogata, T.; Ora, H.; Tanaka, N.; Miyake, Y. Estimation of stride-by-stride spatial gait parameters using inertial measurement unit attached to the shank with inverted pendulum model. *Sci. Rep.* **2021**, *11*, 1391. [CrossRef]
52. Kim, J.; Xie, L.; Cho, S. Comparison of 3D ankle kinematics between minimal inertial measurement units configuration and optical motion capture system under diverse walking conditions. *Sci. Rep.* **2026**, *16*, 8307. [CrossRef]
53. Prisco, G.; Pirozzi, M.A.; Santone, A.; Esposito, F.; Cesarelli, M.; Amato, F.; Donisi, L. Validity of Wearable Inertial Sensors for Gait Analysis: A Systematic Review. *Diagnostics* **2025**, *15*, 36. [CrossRef] [PubMed]

https://doi.org/10.3390/instruments10020023

instruments

MDPI

Review

Electrode Configurations for Electrical Bioimpedance-Based Pulse Wave Signal Acquisition: A Narrative Review

Margus Metshein

Thomas Johann Seebeck Department of Electronics, Tallinn University of Technology, Ehitajate tee 5, 19086 Tallinn, Estonia; margus.metshein@taltech.ee

Abstract

The pulsatile modulation of arterial blood carries essential information about the cardiovascular system and cardiac function—information that can be extracted through appropriate signal-processing algorithms. As wearable technologies are increasingly integrated into everyday life, measurement methods are required to be non-invasive and compact in scale. Electrical bioimpedance (EBI) methods meet these wearability criteria well; however, they introduce uncertainties associated with the electrode–skin interface. This paper presents a targeted overview of electrode configurations for EBI-based pulse wave signal acquisition, focusing on non-invasive solutions suitable for wearable devices. Electrode configurations are examined with respect to major peripheral arteries in the human body that are accessible at the skin surface and suitable for regional impedance cardiography. The review includes a carefully selected set of references, drawing on both research literature and patent descriptions, and discusses the primary differences in how electrode configurations are presented across these sources.

Keywords: brachial artery; carotid artery; electrical bioimpedance; electrodes; hemodynamic feature points; impedance cardiography; pulse wave; radial artery; ulnar artery

1. Introduction

Diseases related to, or originating from, the cardiovascular system are the leading cause of death worldwide. According to the World Health Organization [1], in 2022 cardiovascular diseases (CVDs) accounted for 22% of all deaths and 38% of all premature deaths (under the age of 70). In light of the latter, technology plays a crucial role in the prevention of CVDs through the early detection of emerging disease indicators. Wearable, non-invasive, and unobtrusive monitoring of hemodynamics represents a well-suited application for electrical bioimpedance (EBI)-based measurement devices and methods [2].

Pulsatile blood flow, originating from the heart and propagating throughout the body along the arterial tree, carries essential information about the cardiovascular system as a whole. Information related to the condition of the heart and its ventricles, arterial diseases, arrhythmias, and other pathologies can be expected to manifest as modulations in blood flow [3]. However, the cardiovascular system is a complex entity, and the interpretation of its functions and parameters must take into account the influence of multiple physiological, hemodynamic, and other contributing factors.

While several comprehensive reviews of EBI-based sensing solutions have been published [2,4–7], this paper aims to provide a more narrowly focused overview of EBI-based hemodynamic data acquisition using non-invasive approaches. The emphasis of the present review is on regionally targeted EBI-based electrode configurations and therefore follows the principles of miniaturization, minimal contact with the subject, and intelligent signal acquisition through careful and robust design. The main aim in simple words is the combination of these two aspects: the electrode configuration (1) and arterial site (2).

Within the context of current efforts toward early detection and even prediction of diseases, continuous measurements are increasingly required. Such measurements are expected to be implemented using non-invasive means, ideally unobtrusive to the subject while still being performed ubiquitously. Sensors whose operating principles meet these requirements can be found across a wide range of technologies [8]. However, all such sensors are, in essence, transducers that convert the measured physiological variable into an electrical signal. This signal, typically analog in nature, is digitized in modern wearable systems, which are almost entirely centered around digital signal processing.

The present review focuses on regional impedance cardiography (RICG) in the larger impedance cardiography (ICG) framework, specifically analyzing electrode solutions and configurations targeted at wearable monitoring devices. However, whole-body or thorax-based approaches are not entirely excluded, as such methods have also been integrated into wearable devices [9]. Relevant literature describing EBI measurement techniques has been collected, contextualized, and analyzed. To broaden the review toward practical implementation of EBI measurements, a number of selected patents have also been reviewed and analyzed in parallel within a comparable framework. To the best of the author's knowledge, no similarly focused overview of EBI electrode setups—illustrated through comparative examples from both the scientific literature and patent descriptions—currently exists.

1.1. Narrative Review Literature Source Selection Criterion

The research question for the current review is formulated as follows: "What are the optimal electrode configurations in the combination of arterial sites for wearable EBI-based pulse wave signal acquisition?". While the narrative review in its essence is less structured than the systematic review, often subjective and with inherent bias, it has value in summarizing a very broad chosen topic [10]. Electrode configuration is one of the frame topics for EBI measurements—applicable in all EBI measurement-related applications. In the case of such a broad topic, the targeted information may be present in wide variety of research papers under a number of targeted applications: hemodynamics, plethysmography, impedance cardiography, blood pressure estimation, etc. For this reason, the narrative review was chosen—an opportunity to relay the author's experience in EBI measurement-based wearable application development, irrespective of the area of application.

The criteria for publication date was not defined in searching the sources, as the basics of EBI measurements apply universally (from the past to today). The sources were circumscribed by peer reviewed research papers and conference publications, while no grey literature was included.

The three primary categories for source selection are described as follows. Firstly, the review targets EBI measurement-based human cardiovascular system dynamics monitoring, irrespective of the exact application, and excludes respiratory activity-focused solutions. Secondly, the focus was set on regional body areas, e.g., peripheral arterial sites and limbs. Thirdly, the electrode configurations in the included sources are required to fit into the wearable framework, i.e., be applicable for non-invasive EBI measurements. The third requirement relies on the experience of the author.

The three categories, defined in the section above, contain the following keywords:

- Includes: Bioimpedance, EBI, BioZ, cardiography, and ICG;
- Includes: Pulse wave and artery;
- Includes: Electrode and sensor;

https://doi.org/10.3390/instruments10020026

- Excludes: Respiratory and breathing.

An additional criterion was the presence of numerical $\Delta Z(t)$ values in Ohms in the reference (either specifically stated in numbers or available in plots illustrating the $\Delta Z(t)$). However, example references without numerical expression of $\Delta Z(t)$ values in Ohms were selected to be added into the current review in the cases where the electrode configuration illustrated a general approach.

Using these parameters, the search was conducted across a variety of databases, including PubMed and Scopus; publisher platforms such as IEEE Xplore, MDPI, Springer Nature, Frontiers, Elsevier, IOPscience and Wiley, as well as the Google web search engine.

1.2. Principle of Electrical Impedance

Electrical impedance is defined as the relationship between applied voltage and resulting current, and it is applicable to both direct current (DC) and alternating current (AC) conditions. Under DC excitation, the opposition to current flow is governed solely by resistive effects, which arise from the interaction of charge carriers with the material's internal structure. In contrast, under AC excitation, current flow is influenced by both resistive and reactive components. The reactive contribution originates from induced electromagnetic effects, including magnetic field-related inductance and the capacitive storage of electric charge within the material. The effect of reactance on current flow is comparable to that of resistance; however, unlike resistance, reactance introduces a dependence on signal frequency

$$X = -1/2\pi f C, \tag{1}$$

where X is the reactance, f is the frequency, and C is the capacitance.

In an ideal capacitive element, the voltage developed across it due to the flowing current is phase-shifted by 90°, in contrast to an ideal resistive element where voltage and current are in phase. Specifically, in a capacitor, the voltage lags behind the current as a result of the continuous charging and discharging of its plates. In circuits containing both resistive and capacitive components, the resulting phase-shift lies between 0° and 90°, depending on the relative magnitudes of these elements. Electrical impedance is deduced from Ohm's law and can be expressed by

$$Z = R + jX = ReZ + jImZ, \tag{2}$$

where R is the resistance, Z is the impedance magnitude, and j is the imaginary unit ($j = \sqrt{-1}$). As shown, the impedance Z can be decomposed into its real component (ReZ) and imaginary component (ImZ). Z can be represented as a vector: vector $|\dot{Z}|$

$$|\dot{Z}| = \sqrt{ReZ^2 + ImZ^2}. \tag{3}$$

Impedance magnitude represents the total opposition to current which is the vector sum of resistance and reactance

$$|\dot{Z}| = \sqrt{R^2 + X^2}, \tag{4}$$

with the phase angle, given as:

$$\varphi = arctan(ImZ/ReZ). \tag{5}$$

The same applies to EBI as well, indicating the ratio between the voltage and current by representing the opposition of biological matter to electrical current flow [11]. Biological objects exhibit complex structural characteristics, which in turn lead to increased complexity in the equivalent circuits used for their modeling.

https://doi.org/10.3390/instruments10020026

The primary electrode options for impedance measurements are two- and four-electrode configurations. In the two-electrode arrangement, the measured voltage (V_m) and its resulting measured impedance (Z_m) is not limited to the impedance of interest ($\dot{Z}$), but also includes contributions from the electrode impedances Z_{e1} and Z_{e2}, which are connected in series with the object (Figure 1a). The following can be expressed by

$$Z_m = V_m / I_m = Z_{e1} + Z_{e2} + \dot{Z}, \tag{6}$$

revealing that electrode polarization impedances can introduce significant errors when accurate impedance measurements are required. However, for relative measurements of dynamic biological processes, a two-electrode configuration may still be appropriate due to its robustness and simplicity. For this approach to provide reliable estimates of the true impedance, it is essential that the electrode impedances Z_{e1} and Z_{e2} remain substantially smaller than the impedance of the measured object $\dot{Z}$.

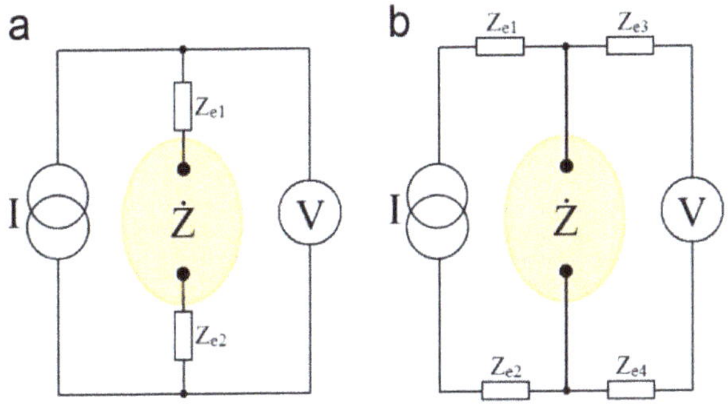

Figure 1. Common impedance measurement systems: two (**a**) and four electrodes (**b**) (where I denotes the excitation signal source, V represents the measured response, and Z_{e1}–Z_{e4} mark the electrode impedances) [12].

The use of separate electrodes for current injection and voltage sensing, commonly referred to as a four-electrode (tetrapolar) configuration, provides an effective means of mitigating the influence of electrode polarization impedance (Figure 1b). In this arrangement, the excitation current is driven through a pair of current-carrying electrodes, while the voltage is measured using a separate pair of sensing electrodes. As a result, the voltage measurement is largely independent of the electrode impedances, and predominantly reflects the impedance of the object under investigation:

$$Z_m = V_m / I_m = \dot{Z}. \tag{7}$$

This holds under the condition that the input impedance of the measurement circuitry is sufficiently high.

In a four-electrode configuration, it is more appropriate to refer to the measured quantity as transimpedance rather than impedance. This distinction arises because the system effectively consists of two independent electrode pairs: one dedicated to current injection and the other to voltage sensing, thereby yielding a transfer characteristic rather than a direct impedance measurement [11]. Furthermore, although this configuration reduces the influence of electrode polarization effects, it does not completely eliminate contributions from lead impedances, sensing electrodes, or artifacts associated with the intrinsic heterogeneity of biological tissues.

https://doi.org/10.3390/instruments10020026

1.3. Hemodynamic Feature Point Extraction—Relevance and Coherence

The impedance cardiography (ICG) waveform, acquired in situ from the thorax using standard electrode positions, exhibits a characteristic pattern that represents the cardiac cycle and the associated volume changes [13]. ICG is the negative first time derivative (dZ/dt) of Z. The exact shape of this pattern depends on the site of acquisition, that is, the nearby artery, and reflects the condition and functional status of the heart and circulatory system as modulated by the blood flow path. Nevertheless, a distinctive waveform with identifiable feature points can be consistently distinguished in the impedance cardiogram.

Moreover, because hemodynamic feature points correspond to specific events occurring during the cardiac cycle, a variety of time intervals can be derived from the impedance cardiogram. For example, the interval between the onset of the Q wave in the electrocardiogram (ECG) (representing depolarization of the interventricular septum in QRS complex) and the B point in the impedance cardiogram—known as the pre-ejection period (PEP)—can be determined (Figure 2) [14].

In practice, the identification of hemodynamic feature (or fiducial) points is fundamentally a signal-processing task. To extract the desired information, the signal must be filtered and conditioned and then converted from analog to digital form, raising considerations related to sampling. Subsequently, the digitized signal is processed—either in real time or offline—to extract the relevant features.

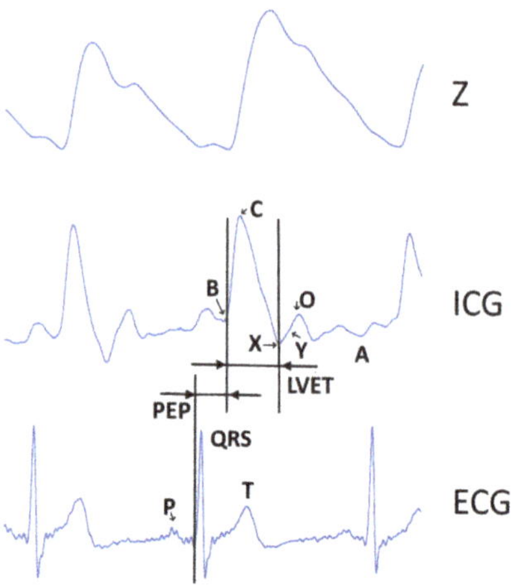

Figure 2. Typical patterns of impedance Z (reflecting volume changes in the body), the corresponding negative 1*st* derivative of Z—the impedance cardiography (ICG) waveform (with hemodynamic fiducial points B, C, X, Y, O, and A)—and the electrocardiography (ECG) waveform (showing the P-wave, QRS complex and T-wave), together with the cardiac time intervals like left ventricular ejection time (LVET) and pre-ejection period (PEP)) (from top to bottom) [15].

In essence, all hemodynamic feature points in the ICG signal correspond to specific cardiac events, that is, changes in conductivity caused by variations in blood volume, which are further influenced by respiratory activity through changes in the amount of non-conductive air in the lungs, body movements, and other factors [16]. These feature points form the basis for a variety of timing calculations, such as pulse transit time (PTT), within a process generally referred to as signal waveform analysis.

The signal itself has substantial intrinsic value, challenging existing methodologies and devices to reliably access the desired information. This raises questions regarding the stage in the measurement chain at which signal preconditioning and feature extraction

https://doi.org/10.3390/instruments10020026

should be performed—either through digital signal processing in the master device or directly within the sensor itself.

Hemodynamic analysis and feature point extraction are important for the detection and treatment of hypertension. Studies have shown that the ICG waveform pattern differs between patients with controlled and uncontrolled hypertension [17]. A focused overview is provided in [18], specifically addressing the prediction and management of hypertension based on hemodynamic data processing.

1.4. Source of the Signal—Volumetric Change of Blood

A simplified depiction of volumetric changes in the aorta and arteries illustrates the rhythmic arrival of pulse waves that cause expansion of the arterial wall diameter (Figure 3 [19]). With the arrival of each pulse wave, the blood volume rhythmically increases and decreases, resulting in corresponding variations in the electrical conductivity of the body volume.

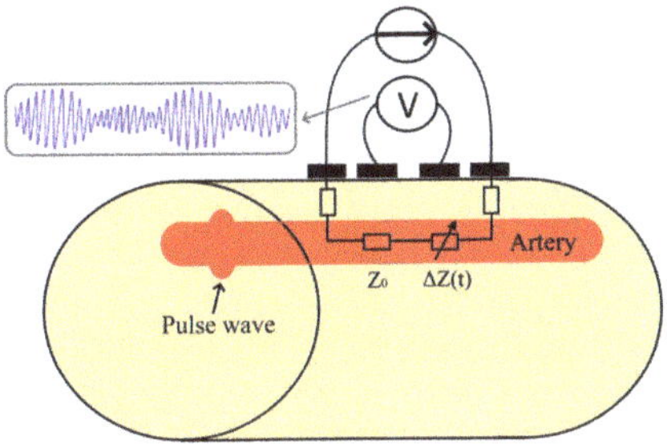

Figure 3. Simplified representation of pulse wave formation in the radial artery including the respective imaginary occurrence of impedance of a local tissue volume Z_0 and a variable component $\Delta Z(t)$ together with the four-electrode EBI measurement setup in which the excitation current (indicated by the arrow) is applied via a dedicated electrode pair and the voltage (V) is measured using a separate electrode pair [20].

This simplified approach relies on the assumption that an electrical current preferentially flows through arterial blood, which has relatively high conductivity (0.703 S/m at 100 kHz [21]) compared, for example, to body fat (0.024 S/m at 100 kHz [21]). Accordingly, the measured impedance Z is considered to consist of an invariant component related to the local tissue volume (Z_0) and a variable component, ($\Delta Z(t)$), caused by the pulsatile flow of blood (which is in essence a volume change) (Figure 4) [20,22]. This classical approach is formulated as follows:

$$Z(t) = Z_0(t) + \Delta Z(t). \tag{8}$$

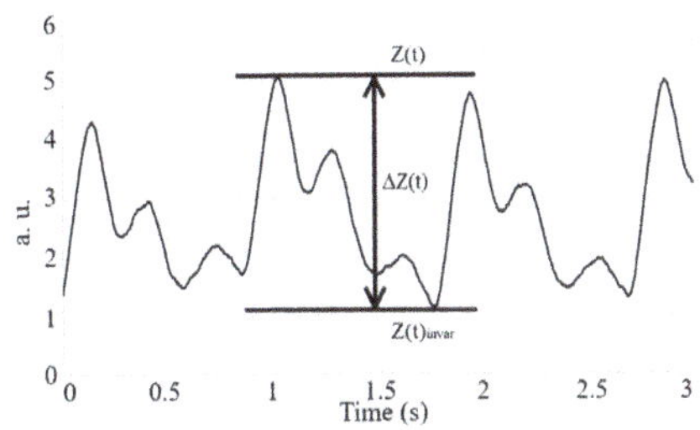

Figure 4. Principle of determining the time varying impedance $\Delta Z(t)$ in the EBI waveform, where the abbreviation a.u. (arbitrary unit) denotes the relative change in volume, as the actual measured value of Z might be misleading in interpreting the waveform (modified from [22]).

https://doi.org/10.3390/instruments10020026

However, in practice—particularly in the case of non-invasive measurements—the effects of multiple tissue layers must be considered, including skin, muscle, bone, tendons, arterial walls, and other structures. These layers exhibit different electrical conductivities: some are only moderately lower than that of blood, such as muscle (0.362 S/m at 100 kHz) or the arterial wall (0.317 S/m at 100 kHz [21]), while others are significantly lower, such as bone (0.021 S/m at 100 kHz [21]). This highlights that changes in the non-invasively measured impedance Z do not arise exclusively from the increased blood volume associated with the propagating pulse wave.

But what is the contribution of blood to the measured impedance? It is well known that the resistivity of blood depends on its flow velocity, a phenomenon known as the Sigman effect [23]. Importantly, the majority of blood resistance arises from its cellular components, particularly red blood cells (RBCs), rather than from the surrounding conductive medium (plasma) [24]. Furthermore, blood is electrically anisotropic; that is, the orientation of RBCs affects blood resistivity and introduces a directional dependence relative to the measurement orientation.

Contact electrodes establish a direct electrical interface with the biological object, which can be achieved either through the use of conductive gel (wet electrodes) or via dry contact. In the latter case, the interface may effectively transition toward a wet condition as the stratum corneum becomes hydrated by perspiration. Alternatively, partial abrasion of the stratum corneum may be employed to improve contact stability, reduce settling time associated with hydration, and minimize the impact of motion artifacts.

In resistive coupling, charge transfer occurs through reversible electrochemical (oxidation–reduction) reactions at the electrode–electrolyte interface, particularly when galvanic contact is established with body fluids such as sweat. This mechanism results in a relatively low-impedance electrical connection between the electrode and the tissue. The interface can be described using established models that incorporate half-cell potentials, double-layer capacitance, and both series and parallel resistive components.

For contact electrodes, the primary objective is to achieve a stable and low-impedance electrical interface with the biological tissue. In clinical practice, this is typically accomplished by applying conductive gels at the skin–electrode interface to enhance conductivity and measurement reliability.

The use of resistive coupling permits operation at relatively low excitation frequencies, as it allows current to pass through the skin–electrode interface. However, this condition is not constant, since the electrical properties of the stratum corneum vary over time and depend on factors such as ambient temperature and psychophysiological conditions. In dry states, the stratum corneum tends to exhibit predominantly capacitive behavior, whereas in hydrated conditions its response becomes more resistive [25].

Under these conditions, individual physiological variations, particularly those related to skin properties, can have a more pronounced impact on measurement accuracy compared to capacitive or inductive coupling approaches. The intrinsic dryness of the skin influences EBI measurements; in such cases, hydration of the skin may be required to ensure reliable measurements when using resistive coupling.

The theoretical analysis presented in [12] suggests that modulation of the measured EBI signal cannot be attributed solely to the pulsatile flow of blood within the artery. The calculations indicate a potentially substantial contribution from muscular tissue, a finding that is supported by results from practical EBI measurements. It can therefore be assumed that the modulation observed in the measured impedance Z originates primarily from the rhythmic compression of surrounding tissues induced by the pulsating arterial blood volume, rather than exclusively from the pulsating blood volume itself.

https://doi.org/10.3390/instruments10020026

When considering non-invasive EBI measurements of the thorax, the situation becomes substantially more complex. The thorax contains a majority of vital organs, a variety of tissues with differing electrical conductivities, and arteries and veins of various diameters carrying large volumes of moving blood. Finite Element Method (FEM) simulations have produced results that challenge the conventional understanding of how different volume changes contribute to the measured thoracic EBI signal [26,27].

The spot-electrode ICG monitoring method proposed by Sramek [28]—derived from a modified version of Kubicek's classical band-electrode approach [29]—has been shown to produce measurable sensitivity distributions in the cardiac region [26]. However, these results are reported to be highly sensitive to electrode placement, with sensitivity primarily concentrated in the upper thorax (above the lungs) and largely dominated by contributions from skeletal muscle [26,27]. In the context of ICG, it has even been speculated that the contribution of the aorta to the total thoracic impedance is less than 1% [27].

When electrodes are targeted to a specific artery, the situation becomes somewhat less complex. In this case, the volumes of different biological tissues involved are smaller, and the resulting sensitivity distribution is more localized [30].

1.5. Electrode-Skin Interface and Motion Artifact

In solid conductive materials, such as electrodes and connecting leads, electrical current is primarily carried by electrons, whereas in electrolytic media, including biological tissues, current conduction occurs via ionic charge carriers [11]. In the context of measurements aimed at characterizing biological systems, an electrode can be defined as the interface at which the transition between electronic and ionic conduction takes place.

When an electrode is brought into galvanic contact with an electrolytic medium, electrochemical reactions, specifically oxidation and reduction processes, occur at the interface. As a result, a half-cell potential (also referred to as the Nernst potential) is established between the electrode and the electrolyte. This potential is material-dependent, as different electrode materials exhibit distinct characteristic values, which can significantly influence measurements, particularly in the low-frequency and direct current (DC) regimes [31].

In the absence of an applied current in a measurement system—typically involving at least two electrodes for signal acquisition or current injection—the observed potential corresponds to the combined equilibrium potentials of the electrode–electrolyte interfaces. When an external current is introduced, these potentials are modified, resulting in a deviation from the equilibrium state. The difference between the zero-current and current-driven conditions affects the distribution of charge within the electrolyte that is in galvanic contact with the electrodes.

Under equilibrium conditions, the electrode potential E is equal to its reversible potential E_{rev}. However, for a real electrode–electrolyte interface, the potential E can be described by the following expression

$$E = E_{rev} + i_{DC} R_{shunt}, \tag{9}$$

where i_{DC} denotes the faradaic current applied at the electrode–electrolyte interface, and R_{shunt} represents the shunt resistance of the interface, which is connected in parallel with the double-layer capacitance.

This phenomenon is referred to as electrode polarization and has a significant impact on electrode performance, highlighting the importance of selecting appropriate electrode materials depending on the measurement conditions. The polarization impedance at the electrode–electrolyte interface is influenced by multiple factors, including frequency, electrolyte concentration, applied voltage, and electrode material, which contributes to the challenges associated with low-frequency dielectric measurements. The extent of

https://doi.org/10.3390/instruments10020026

polarization can be controlled through material selection; for example, noble metals such as platinum, silver, and gold exhibit minimal chemical reactivity with biological fluids and typically do not support substantial charge transfer, thereby demonstrating predominantly polarizable behavior.

Electrode polarization can be attributed to the formation of an electrical double layer at the electrode–electrolyte interface during charge transfer between electronic and ionic carriers. For charge transfer to occur across this interface, ions must migrate toward the electrode under the influence of electrostatic forces. This process leads to the accumulation of oppositely charged layers on either side of the interface, resulting in a layer of positive charge at the electrode surface and a corresponding layer of negative charge within the adjacent electrolyte [32].

When an electrode is brought into contact with an electrolyte solution, an electrical double layer forms almost instantaneously at the interface. This structure can be approximated as a parallel-plate capacitor, where the effective separation between the charge layers is on the order of 0.1–10 nm, resulting in relatively large capacitance values [11].

A critical aspect in the monitoring of electrical bioimpedance (EBI) in biological systems is the presence of motion artifacts [33]. These artifacts arise from various sources, including movements at the electrode–skin interface as well as internal movements within the biological object itself. Similar to other types of interference, motion artifacts can be regarded as noise that distorts the measured EBI signal or other biopotentials by introducing additional unwanted voltage components. In particular, motion at the electrode–skin interface may result from displacement of the electrodes relative to the surface of the subject, as well as from relative movement between the layers of the skin. Although motion artifacts are especially pronounced when using capacitive electrodes, they are also present in measurements performed with conventional low-resistance (wet) contact electrodes.

Electrode motion can generally be classified into two categories: transversal (normal to the skin surface) and lateral (parallel to the skin surface) [34]. Transversal motion alters the skin–electrode contact impedance by changing the distance between the electrode and the skin, leading to variations in polarization at the interface and the generation of additional voltage components. In contrast, lateral motion, particularly in direct-contact configurations, can induce friction at the interface, resulting in triboelectric charge accumulation that contaminates the measured signal.

To mitigate the effects of motion artifacts, conductive gels are commonly applied at the skin–electrode interface to reduce polarization impedance. In addition, various complementary approaches have been proposed, including signal processing techniques such as adaptive filtering [35], as well as motion compensation methods based on auxiliary sensors, for example, accelerometers attached to the electrode system [36].

Relative motion between different layers of biological tissue, such as between the dermis and epidermis, can give rise to skin potential artifacts. These artifacts may reach amplitudes of several millivolts, depending on the degree of mechanical deformation, including stretching and applied pressure. Partial mitigation can be achieved by reducing the influence of the stratum corneum. In addition, improvements in electrode design have been introduced to minimize these artifacts. For instance, incorporating a conductive foam layer between the electrode and the skin can be effective. Owing to its elastic properties, the foam helps maintain stable contact during movement while minimizing friction at the interface [37].

1.6. Basics of Electrode Placement for EBI Measurement

The positioning of non-invasive electrodes relative to the object of interest, located beneath layers of tissue with unknown or highly variable properties (such as skin), constitutes a multifaceted challenge. If the object of interest occupies a sufficiently large volume relative to the total measurement volume, or is located deep within the body (e.g., the

https://doi.org/10.3390/instruments10020026

heart), electrodes are traditionally placed at the extremities of the body. Such approaches are well illustrated in the work of W. G. Kubicek et al. [29] and B. B. Sramek [28].

When the object of interest—or the associated volume change—within the body is small relative to the total body volume, it is appropriate to target the electrodes directly toward that object (e.g., [28] or [19]). In hemodynamic parameter monitoring, such an object may be the blood within vessels, that is, the pulsatile blood volume in an artery. When focusing on a relatively small structure located close to the skin surface, the electrodes can be placed in close proximity to one another.

A classical and intuitive approach to electrode placement for detecting the dynamics and properties of arterial blood flow is to arrange the electrodes "in a row along the artery" (Figure 5a). In this configuration, the electrodes are positioned above the artery, with the expectation that the injected current penetrates the intervening biological layers and interacts with the blood volume, as illustrated in Figure 3. As the current is assumed to flow predominantly along the pulsating blood volume within the artery, its contribution to the measured signal is expected to be maximal. Importantly, in the majority of cases, the excitation signal is applied through the outermost electrodes, while the response signal is acquired (measured) using the inner pair of electrodes (Figure 5a).

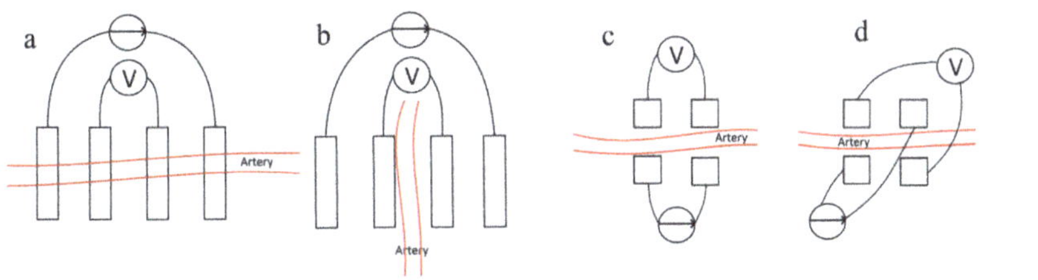

Figure 5. Classical four-electrode configurations with electrodes arranged both along (**a**) and across the artery (**b**) as well as less common rectangular (**c**) and perpendicularly oriented electrode placements (**d**) (where the arrow indicates the application of the excitation signal and V denotes the acquisition of the response signal).

An important consideration is the selection of electrode configuration, with the most common options being two- and four-electrode setups (a thorough overview can be found in [38]). A fundamental difference exists between these configurations: in a two-electrode setup, only positive sensitivity is present, whereas in a four-electrode setup, the perpendicular intersection of current density vectors can result in regions of negative sensitivity [11]. According to theory, negative sensitivity corresponds to a decrease in the measured impedance Z [11].

Considering the layered structure of the human body (e.g., the forearm), individual sub-volumes can exert complex effects on EBI measurements, particularly in the case of four-electrode configurations [39].

However, alternative electrode placements exist, reflecting the complexity of the human body and the resulting uncertainty in current density distribution within its volume [30]. A simple modification of the "in a row along the artery" approach is to rotate the electrode configuration by 90 degrees relative to the direction of the artery (Figure 5b). In this configuration, the artery is positioned between the measuring electrodes. Although this approach is commonly used in practical applications, it has been reported only sparingly in the scientific literature [20,40].

Other non-standard electrode placement strategies also exist, based on the assumption that sensitivity distributions within the human body are considerably more complex. These approaches assume that the regions of highest sensitivity do not form directly beneath the electrodes. Examples are the rectangular and perpendicular electrode placements relative

https://doi.org/10.3390/instruments10020026

to the artery (Figure 5c,d) [30]. A representative selection of such electrode configurations has been presented, simulated, and experimentally evaluated in [30].

Efforts in using electrical impedance tomography (EIT) in pulse wave-related parameter estimation are described in the literature. EIT is a non-invasive imaging technique that provides real-time, cross-sectional images of electrical conductivity within the body. The primary target in applying EIT has been the central aortic blood pressure [41,42]. The idea behind the usage of EIT has been the visualization of certain cardiac events like systole—the ejection of blood into the arteries [41] or cardiac output [43]. Although the EIT-based pulse wave acquisition solutions are still in their infancy due to the imaging accuracy, the topic itself is vastly wide and the current paper is mostly dedicated to focusing on the volume change itself. Hence, EIT solutions are not discussed further in the current paper.

2. Pulse Waveform and Its Interpretation Through ICG

The shape of the pulse wave differs depending on the site of acquisition [44]; that is, the characteristic pattern of the detected pulse waveform varies among the aorta, carotid, brachial, and radial arteries (Figure 6 [44]).

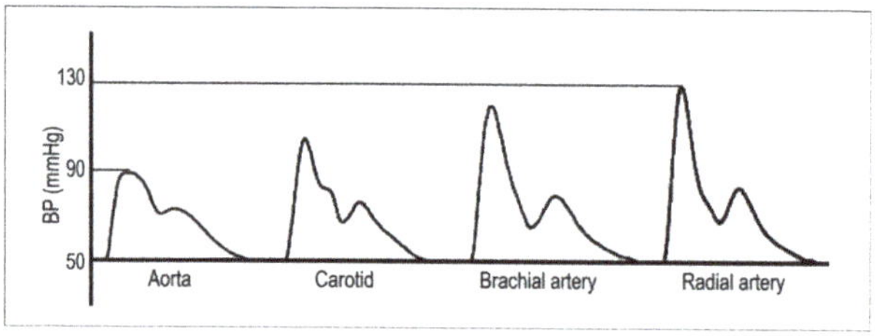

Figure 6. Evolution of the blood pressure waveform as a function of distance from the heart to the peripheral arteries [44].

The waveform can be characterized by distinctive markers that represent key events in blood flow dynamics. The following key points have been defined for the arterial pressure waveform, obtained, for example, by applanation tonometry (Figure 7):

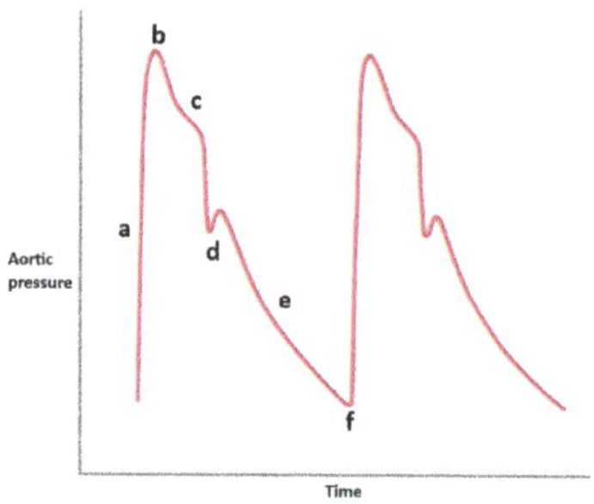

Figure 7. Arterial pressure waveform showing the following characteristic points: systolic upstroke (a), peak systolic pressure (b), systolic decline (c), dicrotic notch (d), dicrotic runoff (e), and end-diastolic pressure (f).

Similarly, the EBI signal (and, consequently, the ICG signal) varies depending on the acquisition approach, that is, the specific electrode configuration and placement on the body surface [45]. In general, the real part Z of the impedance signal exhibits a characteristic double-trough pattern. The first trough corresponds to the arrival of the blood pulse at the electrode location, representing systole. The second trough represents the reflection of the pulse wave from the peripheral vascular bed, such as the capillaries in the fingers and toes [46].

https://doi.org/10.3390/instruments10020026

The ICG, which represents the time derivative of the impedance signal Z (dZ/dt), is characterized by a set of predefined hemodynamic feature points and time intervals (Figure 2):

- A—Atrial contraction;
- B—Opening of the aortic valve;
- C—Maximum systolic flow;
- X—Closing of the aortic valve;
- Y—Closing of the pulmonary valve;
- O—Opening of the mitral valve;
- PEP—Pre-ejection period;
- LVET—Left ventricular ejection time.

Importantly, ICG reflects the mechanical activity of the heart, whereas ECG represents its electrical activity [11]. Consequently, the onset of the QRS complex in the ECG signal always precedes the B point in the ICG signal. As the site of impedance Z measurement moves farther from the aorta, the time interval between the commonly monitored R wave of the ECG QRS complex and the B point of the ICG signal increases (Figure 2). Moreover, the waveform pattern itself varies depending on the exact sensing location [45].

3. Electrode Positioning and Locations for ICG Signal Acquisition

In the literature, the methods for ICG signal acquisition have been divided into three categories (Figure 8) [47]:

- Whole-body ICG (WICG);
- Thoracic ICG (TICG);
- Regional ICG (RICG).

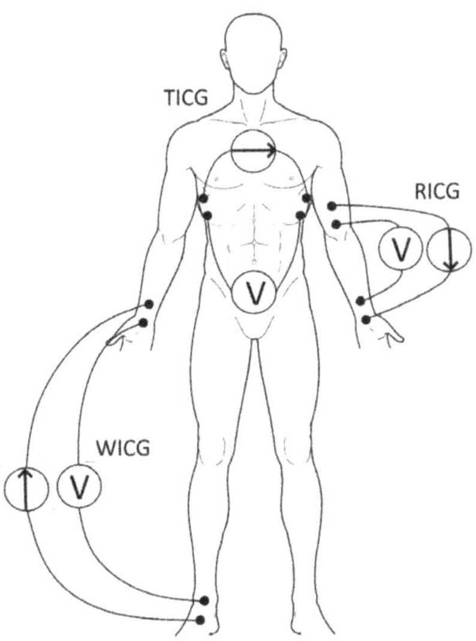

Figure 8. Impedance cardiography (ICG) signal acquisition methods: whole-body ICG (WICG), thoracic ICG (TICG), and regional ICG (RICG), where the arrow indicates the application of the excitation signal and V denotes the acquisition of the response signal.

With certain limitations, all three approaches—WICG, TICG, and RICG—can be considered suitable for wearable monitoring device design. However, the RICG approach is the most convenient, as it enables compact and miniaturized device designs through dense electrode

https://doi.org/10.3390/instruments10020026

arrangements. The choice of electrode placement strategy for acquiring the EBI signal (and its transformation into ICG) depends strongly on the targeted hemodynamic parameter(s).

For example, the approach may involve a standard procedure such as venous occlusion plethysmography to assess venous volume for the detection of arterial and venous disease [48]. Alternatively, it may involve blood pressure estimation, which requires determination of pulse wave velocity (PWV) at two distances from the heart (as demonstrated in [49,50]). This latter approach is not yet a standardized procedure and can be implemented using measurements from various anatomical locations.

3.1. ICG Signal Acquisition—Whole-Body Locations

The WICG approach encompasses configurations in which electrodes are attached to two or more limbs, classically in a hand–foot arrangement, though not necessarily limited to this configuration. Examples include the classical four-electrode wrist–ankle measurement setup [51] or a corresponding two-electrode alternative [52]. This paper does not focus on addressing the WICG approach in detail.

3.2. ICG Signal Acquisition—Thoracic Locations

TICG is the original ICG signal acquisition method, in which electrodes are attached at the boundaries or at predefined, historically established positions on the thoracic area [29]. However, more targeted and focused approaches have also been proposed [53,54].

As defined in [11], "ICG is impedance plethysmography based on the measurement of thoracic EBI". The thorax is a classical site for acquiring the ICG signal, a practice pioneered by W. G. Kubicek et al. in the 1960s [29]. However, the method of impedance plethysmography itself was formulated earlier by J. Nyboer et al. in 1950 for cylindrical body regions such as the arms and legs [55]. The corresponding formula was derived to estimate increases in blood volume ΔV from the simultaneous decrease in the measured EBI time series (i.e., from the depth of modulation in the EBI signal—ΔZ)

$$\triangle V = -(\rho L^2/Z^2)\triangle Z, \tag{10}$$

where ρ is the resistivity, L is the length of the cylindrical body segment, and Z is the mean value of the measured impedance. Here, the term "cylindrical" highlights an inherent limitation, as most parts of the human body are not truly cylindrical.

In essence, ICG signal acquisition involves measuring the thoracic EBI using appropriate electrode placement strategies and predefined excitation signal parameters. The first impedance cardiograph employing electrodes was developed by R. Patterson et al., utilizing band electrodes in a four-electrode configuration [13]. The electrodes were attached around the trunk and neck to reduce the effects of patient motion. However, the limitations of band electrodes soon began to hinder the further development of the ICG method, primarily due to the non-uniform distribution of contact impedances between the band electrodes and the skin surface [13].

As a solution, spot electrodes were proposed by B. B. Sramek to limit the electrode–skin contact area [28]. In addition, the Sramek method sought to address difficulties arising from blood resistivity in volume calculations. This approach involved decomposing the impedance into component terms and introducing a volume-related coefficient representing the electrically participating tissue volume V_{EPT} [56]. A limitation of this method is that it provides only a statistical estimate, as individual blood resistivity is not taken into account [56].

Since then, ICG methods have evolved and become increasingly sophisticated across all modalities, including electrode placement strategies, interpretation of the measured signals, and the development of measurement instrumentation. In addition, modified ICG signal acquisition approaches have been proposed, either targeting specific arteries for

https://doi.org/10.3390/instruments10020026

wearable device applications [54,57,58] or aiming to minimize the number of electrodes by co-utilizing the same electrodes for both ICG and ECG measurements [59].

However, driven by the desire to reduce hardware size and improve wearability, anatomical locations other than the thorax have been explored for ICG measurements [59]. Since the source of volumetric change in the circulatory system is the heart and arterial blood reaches all capillaries, modulation of the EBI signal can also be expected at virtually any arterial site. The current manuscript does not concentrate on TICG electrode configurations, but serves as a reference to RICG aspects.

3.3. ICG Signal Acquisition—Non-Thoracic Locations

RICG encompasses all measurement locations on the body surface that are situated outside the thoracic area, such as a single limb or a specific body region. This approach may focus on blood volume changes carried by a particular artery [45] or on localized tissue volumes, such as specific tissue layers [60] or pathological regions (e.g., cancer [61]). The targeted features or hemodynamic parameters may vary accordingly and often require advanced signal processing due to the low signal amplitude and noisiness of the acquired signal.

For example, PWV estimation is classically performed non-invasively using measurements at the carotid and femoral arteries [62]. Carotid–femoral PWV (cfPWV) is considered a gold-standard method due to its reproducibility and ease of application [63]. To obtain cfPWV, concurrent EBI measurements are performed at the neck and the groin region, with sensitivity depending even on the selection of the specific carotid artery branch [64], typically using a four-electrode configuration.

The wrist is one of the most intuitive locations for ICG signal acquisition. It is a classical site for wristwatches—which have not yet been displaced by other technological solutions—and are easy to attach and operate using the opposite (typically dominant) hand. The wrist has also been a popular measurement location in combination with the thorax in PWV determination for blood pressure estimation [50,65].

However, more recent solutions have reported the implementation of PWV measurements using the wrist alone [49,50]. Such approaches require two simultaneously measuring electrode pairs separated by a known distance to enable velocity calculation based on pulse transit time (PTT) and the distance between electrode sites [50].

Importantly, determination of the PTT requires identification of only the peak systolic pressure, which corresponds to the C point in the ICG signal; that is, only a single feature point needs to be extracted. The C point is the most distinctive feature of the ICG waveform and is classically defined as the maximum peak value of the signal (i.e., the global extremum) [66].

Extraction of other hemodynamic feature points using RICG (including measurements at the wrist) is more challenging. This is because the modulation of arterial blood flow caused by the mechanical activity of the heart is significantly attenuated by the time it reaches the measurement site. Nevertheless, extraction of fiducial points is still possible, as demonstrated, for example, in [58], but it depends directly on the quality of the acquired impedance cardiogram, that is, on the quality of the measured EBI signal.

However, the wrist is not the only peripheral site suitable for impedance monitoring to obtain the ICG signal. A number of arteries that are sufficiently close to the skin surface for non-invasive EBI measurements are present in other parts of the body, including the ulnar artery; the left and right common carotid arteries; the brachial artery; and the anterior and posterior tibial arteries, among others (as demonstrated, e.g., in [45,67]). As shown in Figure 6, the shape of the blood pressure waveform varies depending on the measurement site—and the EBI (but also the photoplethysmoghraphy (PPG)) waveform varies accordingly (Figure 9).

https://doi.org/10.3390/instruments10020026

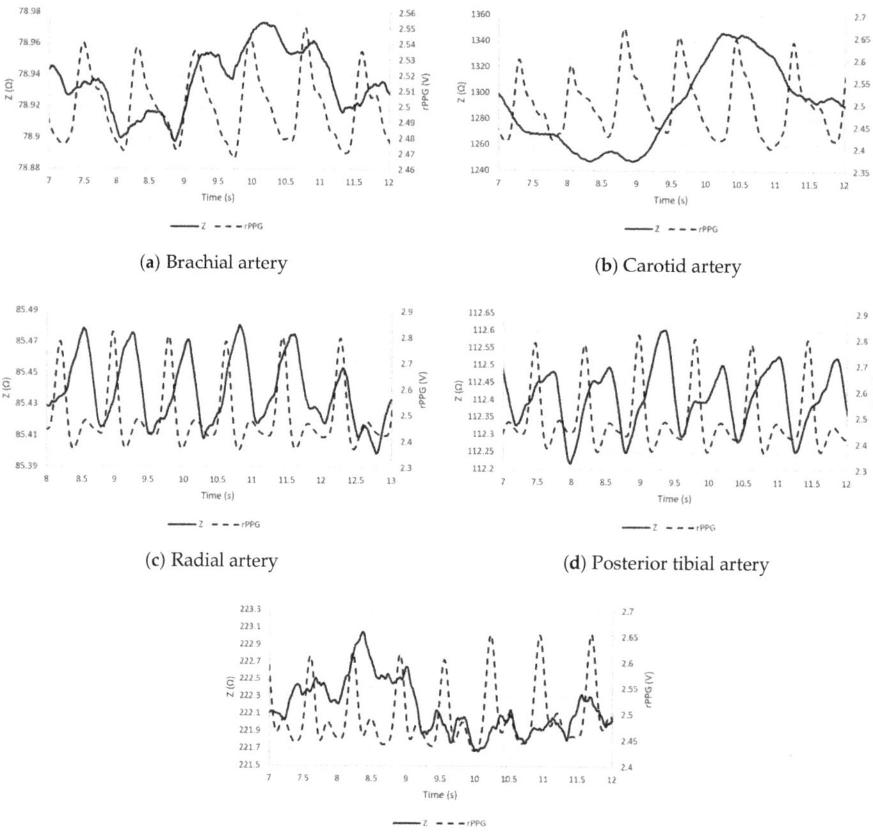

(**a**) Brachial artery

(**b**) Carotid artery

(**c**) Radial artery

(**d**) Posterior tibial artery

(**e**) Ulnar artery

Figure 9. Waveforms of EBI (measured at an excitation frequency of 100 kHz) and reflection mode photoplethysmoghraphy (rPPG) signals acquired from the following selected sensor locations: (**a**) brachial artery; (**b**) carotid artery; (**c**) radial artery; (**d**) posterior tibial artery; and (**e**) ulnar artery [67].

RICG electrode arrangements can be categorized into two types: (1) miniaturized configurations focused on a specific artery within a limb or body region, and (2) configurations that cover an entire limb or body segment. In the first approach, electrodes are typically placed in close proximity, and accurate placement requires knowledge of the anatomical location of the targeted artery (e.g., [19,49,50]). In the second approach, electrodes are generally positioned farther apart, either to avoid the need for precise artery targeting or because the artery of interest lies deeper beneath or between tissue layers (e.g., [65,68]).

An overview of various RICG approaches is presented in Tables 1–3. These tables summarize selected scientific literature that reports EBI measurements on regional body areas targeting specific arteries and compare the corresponding approaches and performance characteristics. The selection was designed to include a range of arterial locations, reflecting their frequency of appearance in the published literature.

An overview of patented RICG approaches is provided in Table 4. This table is not intended to be an exhaustive survey of all patents related to EBI measurement apparatuses and methodologies, but rather presents a representative selection of electrode arrangement approaches.

https://doi.org/10.3390/instruments10020026

Table 1. Selected electrode arrangement solutions based on RICG targeted to radial artery from the literature.

Authors (Year)	Targeted Artery	Electrode Properties	Electrode Arrangement	Features Extracted	Population (Cycles)	Highest and Lowest $\Delta Z(t)$
Namkoong et al. (2023) [69,70]	Radial	Nanocomposite film (Ag) (0.084 cm^2) with a gap of 0.9 cm.	In row along the artery	Signal waveform analysis (BP)	1 (not specified)	(max) 2.7 Ω (min) 0.06 Ω (from graph)
Namkoong et al. (2023) [69]	Radial	Ag/AgCl (0.084 cm^2) with a gap of 0.9 cm.	In row along the artery	Signal waveform analysis (BP)	1 (not specified)	(max) 0.2 Ω (min) 0.07 Ω (from graph)
Huang et al. (2017) [71]	Radial	Copper plated PCB (0.3 cm^2) with gap 0.3 cm	In row along the artery and middle finger	PWV (BP)	10 (not specified)	-
Huynh et al. (2019) [72]	Radial	Silver-plated polyester (0.75 cm^2) with gap 0.5 cm	In row along the artery	PPT (BP)	15 (not specified)	-
Lee and Yoo (2021) [73]	Radial	Dimensions not specified (flexible PCB with wet-contact gel-based EMS sheet)	In row along the artery	PWV, PTT (BP)	3 (not specified)	-
Ibrahim and Jafari (2022) [74]	Radial	Silver (0.25 cm^2) with gap 0.32 cm	In row along the artery	PTT (BP)	4 (not specified)	(max) 0.1207 Ω; (min) 0.046 Ω
Anand and Lowe (2020) [75]	Radial	Ambu Whitesensor ECG electrode (of modified size) with 1 cm gap	In row along the artery	Signal waveform analysis	3 (not specified)	(max) ca. 7 Ω (min); ca. 3 Ω (from graph)
Metshein et al. (2018) [19]	Radial	Gold-plated circular PCB with a gap of 1.25 cm between the exc. and 0.5 cm to meas. el.	In row along the artery	Signal waveform analysis (CAP)	1 (3)	(max) 4.637 Ω; (min) 1.384 Ω
Pesti et al. (2021) [30]	Radial	Ag/AgCl modified to 1 cm width with gap 0.5 cm	In row along the artery	Signal waveform analysis (CAP)	1 (3)	(max) 0.15 Ω; (min) 0.036 Ω (from graph)
Ibrahim et al. (2019) [76]	Radial	Ag/AgCl with ca 1 cm width	In row along the artery	PTT (BP)	1 (not specified)	(max) 0.4 Ω; (min) 0.2 Ω (from graph)

https://doi.org/10.3390/instruments10020026

Table 2. Selected electrode arrangement solutions based on RICG targeted to ulnar and radial arteries together with any other arteries from the literature.

Authors (Year)	Targeted Artery	Electrode Properties	Electrode Arrangement	Features Extracted	Population (Cycles)	Highest and Lowest $\Delta Z(t)$
Imbriglia et al. (2024) [77]	Ulnar	Small Ag/AgCl with a gap of 1 cm	In row along the artery in two-electrode setup	PWV (BP)	1 (not specified)	(max) 0.15 Ω; (min) 0.11 Ω (from graph)
Huynh et al. (2018) [50]	Across both: radial and ulnar	3M copper foil (0.25 cm^2) with 2.5 cm gap between the exc. and 1 cm gap to meas. el.	In row along and across the artery	HR, PWV (BP)	15 (not specified)	(max) 0.3258 Ω; (min) 0.0672 Ω
Cho et al. (2009) [49]	Radial/ulnar	Rectangular dry, flexible (0.75 cm^2) with 1 cm gap	In row along the artery	HR, PWV (BP)	1 (43)	Radial (max) 0.1187 Ω (min) 0.0452 Ω; ulnar (max) 0.1311 Ω (min) 0.0481 Ω
Rachim and Chung (2019) [78]	Radial/ulnar	Silver-plated polyester with unspecified gap	In row across the artery	PTT (BP)	10 (not specified)	-
Bang et al. (2009) [79]	Radial/brachial	Dimensions not specified (assumably Ag/AgCl)	In row along the artery	PTT (BP, arterial stiffness)	1 (10)	-
Sel et al. (2021) [45]	Radial/anterior tibial/left common carotid	ECG electrode (of original size) with gaps of 0.5 cm	In row along the artery	Signal waveform analysis	1 (900+)	Radial: R = 0.1 Ω, X = 0.02 Ω; anterior tibial: R = 0.06 Ω, X = 0.015 Ω; carotid: R = 0.25 Ω, X = 0.04 Ω (in average)

https://doi.org/10.3390/instruments10020026

Table 3. Selected electrode arrangement solutions based on RICG targeted to all other arteries from the literature.

Authors (Year)	Targeted Artery	Electrode Properties	Electrode Arrangement	Features Extracted	Population (Cycles)	Highest and Lowest $\Delta Z(t)$
Bernstein et al. (2012) [68]	Brachial	3M, Red Dot TM electrodes with a gap of 5 cm	In row along the artery	B-, C-, X-, point LVET (CO, SV)	29 (not specified)	(max) 0.428 Ω; (min) 0.13 Ω
Liu et al. (2017) [80]	Brachial	Ag/AgCl with a gap of 4 cm	In row along and across the forearm	PTT (BP)	20 (not specified)	(max) ca. 1.5 Ω (min); ca. 0.2 Ω (from graph)
Ibrahim et al. (2017) [81]	Brachial	Not specified (with gaps of 1–3 cm)	In row along the artery	Signal waveform analysis (CAP)	1 (not specified)	(0.3 Ω (in average))

https://doi.org/10.3390/instruments10020026

Table 3. *Cont.*

Authors (Year)	Targeted Artery	Electrode Properties	Electrode Arrangement	Features Extracted	Population (Cycles)	Highest and Lowest $\Delta Z(t)$
Wang et al. (2021) [82]	Carotid	AgCl (variety of dimensions and gaps	In row along the artery	$\Delta Z(t)$ (BP, SV, CAP)	6 (not specified)	(max) 2.137 Ω; (min) 0.895 Ω
Wang et al. (2022) [83]	Carotid	Flexible silver-plated polyester textile (1.8 cm^2) with gap 0.5 cm	In row along the artery	PTT (BP)	6 (not specified)	-
Collette et al. (2011) [84]	Carotid/femoral	ECG electrode (of original size) with 20 cm gap between the exc. and meas. el. pair	In row along the artery	PWV (arterial stiffness)	118 (not specified)	-
Díaz et al. (2010) [85]	Lateral plantar, posterior tibial, medial plantar	Unspecified (commercial bathroom weighing scale with dry electrodes)	In row along the artery, under the foot	HR (not specified)	10 (not specified)	-
Tronstad et al. (2023) [86]	Dorsalis Pedis	Electrode ink with Ag/AgCl (0.5 cm^2) with gap 0.5 cm	In row along the artery	PWV (BP)	1 (not specified)	(max and min) 0.07 Ω

https://doi.org/10.3390/instruments10020026

Table 4. Selected electrode arrangement solutions based on RICG from patent descriptions.

Patent No (Date)	Targeted Artery	Electrode Properties	Electrode Arrangement	Features Extracted	Hemodynamic Parameter
US 10 610 111 (granted 7 April 2020) [87]	Radial	Unspecified dimensions (miniature) and gaps	Not specified	PWV	CV, CO, BP, CHF
US 15/620,651 (priority 22 September 2016, abandoned) [88]	Radial/(ulnar)	Unspecified dimensions (miniature) and gaps	In row along the artery/meas. electrodes shifted to both sides	PWV	HR, HRV, BP
US 11 320 789 (granted 3 May 2022) [89]	Radial/ulnar	Flexible, miniaturized (unspecified dimensions)	In row along the artery	Pulse waveform	HR, BP
US 11 185 246 (granted 30 November 2021) [90]	Radial/ulnar	Unspecified dimensions (miniature) and gaps	Meas. pair on radial artery and exc. pair on ulnar artery	EBI signal	Not specified
US 14/491,836 (priority 19 September 2014, abandoned) [91]	Radial/ulnar	Unspecified dimensions (miniature) and gaps	In row along the artery	IP, pulse waveform	Not specified
US 10 959 622 (granted 30 March 2021) [92]	Forearm (radial/ulnar)	Unspecified dimensions and gaps	In row along the limb	PWV, PTT	BP

https://doi.org/10.3390/instruments10020026

Table 4. *Cont.*

Patent No (Date)	Targeted Artery	Electrode Properties	Electrode Arrangement	Features Extracted	Hemodynamic Parameter
US 11 317 814 (granted 3 May 2022) [93]	Radial/ulnar/intermediate	Unspecified dimensions (miniature) and gaps	In row along the artery	Signal waveform analysis	BP, HR
US 9 603 533 (granted 28 March 2017) [94]	Brachial, carotid, radial, ulnar, femoral, digital	Variety of dimensions	Meas. and exc. pairs on separate sides of artery (crosswise)	PWV	BP
US 10 278 599 (granted 7 May 2019) [95]	Brachial	Unspecified (flexible substrate)	Meas. and exc. pairs on separate sides of artery	PWV, PAT, PTT, vascular transit time (VTT)	SV, CO, BP
US 11 284 809 (granted 29 March 2022) [96]	Any artery in limbs (primarily radial, ulnar, or brachial)	Flexible with variety of gaps and dimensions (0.025–1 mm^2)	In row along the artery	Pulse waveform	HR, BP
US 17/026,929 (priority 21 September 2020) [97]	Carotid, ascending pharyngeal	Unspecified dimensions (miniature) and gaps	Not specified	Pulse waveform	HR
US 9 211 073 (granted 15 December 2015) [98]	Subclavian	Unspecified dimensions and gaps	Meas. and exc. pairs on separate sides of artery	LVET, PTT, pulse waveform	CO, SV, BP, HR
US 62/346,410 (priority 6 June 2016) [99]	Aortic arc, subclavian	Unspecified dimensions and gaps	Meas. and exc. pairs on both sides of subclavian arteries	PAT, VTT, signal waveform analysis	BP, HR, HRV, SV, CO
US 16/478,848 (priority 24 January 2018) [100]	Heart	Variety of dimensions (80–300 mm^2) and gaps (5 mm etc.)	Designed specifically according to the heart location	PWV	BP, CO etc.

https://doi.org/10.3390/instruments10020026

4. Effect of Electrode Geometry and Distance on the Acquired Signal

Changing the dimensions and geometry of the EBI excitation and sensing electrodes, as well as the distance (i.e., gap) between them, affects the acquired ICG signal waveform and its shape. Through miniaturization, electrodes can be adapted into forms that are more suitable for specific locations on the body surface where the target artery is accessible. In a two-electrode system, the inter-electrode distance directly determines the effective measurement depth. In contrast, the situation is considerably more complex in a four-electrode system [11].

By separating the excitation and sensing electrodes, the lead field theory is commonly used as the underlying framework. The lead field theory characterizes the distribution of measurement sensitivity with respect to the excitation and sensing electrodes within a finite volume [101]. However, the accuracy of the resulting sensitivity distribution depends on the complexity of the model, including the specified volume geometries and resistivity distributions [11].

Moreover, in the case of RICG—especially for miniaturized designs—the positioning of electrodes relative to the target volume change (e.g., pulsatile blood flow in the radial artery) is a critical consideration [11]. The effects of electrode displacement in measurements at the radial artery have been demonstrated in the literature, highlighting the importance of a well-defined signal acquisition strategy [19,102]. In general, "in a row" electrode placement directly above the artery yields the best performance in terms of signal-to-noise ratio (SNR) and the magnitude of $\Delta Z(t)$ [19,102].

However, alternative electrode positioning strategies also exist, typically still located above the region of interest where the volume change occurs. For example, non-standard configurations may place excitation electrodes on one side of the artery and sensing electrodes on the opposite side [30]. In such applications, electrode arrays are often employed (e.g., [74,103]), comprising multiple connection sites that can be selectively switched or programmed depending on the quality of the acquired signal. Based on predefined quality criteria—potentially supported by artificial intelligence (AI) techniques—the most suitable electrode configuration can then be selected.

The focused impedance method (FIM) has also attracted increasing interest since its original introduction by Rabbani et al. in [53]. The principle of FIM is to concentrate the regions of excitation current and measured potential onto a specific biological volume conductor or organ by appropriately arranging the electrodes [104], based on lead field theory. The idea is realized by applying two excitation channels with equal parameters and picking up the response voltage (e.g., by using eight electrodes (Figure 10)). A modified version of the FIM, as presented by the authors of [30], has demonstrated promising results for EBI-based pulse wave acquisition from the radial artery at the wrist.

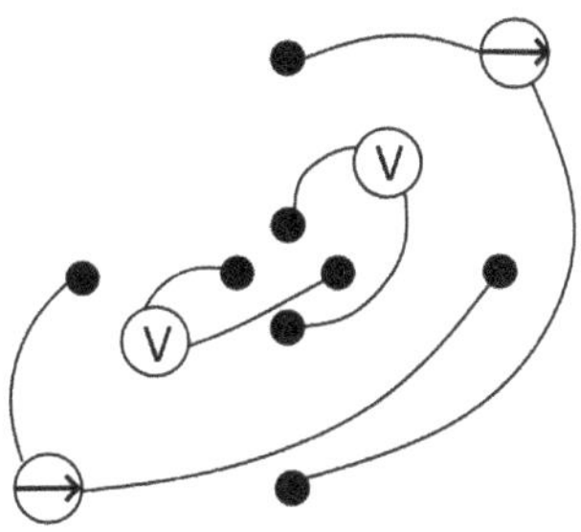

Figure 10. Example of an eight-electrode configuration for the focused impedance method, where the arrow indicates the application of the excitation signal, V denotes the acquisition of the response signal, and the black dots represent the electrodes.

https://doi.org/10.3390/instruments10020026

The sensitivity distribution thus becomes a critical issue, raising questions about the true origin of the measured signal. It is well known that electric current preferentially flows through media with higher electrical conductivity. Consequently, it is often assumed—sometimes oversimplified—that the majority of the current flows through the pulsatile blood volume within arteries [49,79,102,105]. However, depending on muscle type and physiological state, a substantial and highly variable volume of blood also flows through muscle tissue (approximately 5–10 mL/min per 100 g at rest, increasing to 80–100 mL/min per 100 g during maximal vasodilation or active hyperemia) [106].

In non-invasive electrical measurements, blood perfusing muscle tissue can therefore be expected to shunt the pulsatile arterial blood flow. Nevertheless, modulation of the measured EBI waveform is still present, primarily caused by the mechanical pressure exerted by the expanding arterial wall—whose compliance depends on several factors such as aging and PWV—on the surrounding muscle tissue [107].

Simulation studies, particularly FEM modeling using various electrode placement strategies, have revealed a wide range of possible sensitivity distributions, which generally concentrate in the region of the targeted artery. However, relatively few such models have been developed for specific arteries, such as the radial artery, leaving the exact source of the measured signal still under debate [30,108,109].

Another important aspect in this context is contact resistance, whose value increases as electrode dimensions decrease [110]. The impedances of the electrodes and the skin—whose condition is highly sensitive to factors such as air humidity, emotional state, and skin moisture—can differ significantly, leading to deterioration of the SNR. Mismatch in measuring electrode contact impedance is a source of error that can distort the measured impedance Z and reduce measurement sensitivity [111]. To address this issue, various contact impedance compensation schemes have been proposed for wearable devices (e.g., [110]).

The quantitative analysis of the numerical data (Table 5) derived from the literature sources (Tables 1–3) reveals a general trend. Table 5 shows the calculated average value of $\Delta Z(t)$ over a gap between the measuring electrodes, i.e., the value of $\Delta Z(t)$ is divided into the gap between the measuring electrodes in each literature reference and then averaged.

Table 5. Analysis of the average maximum and minimum values of $\Delta Z(t)$ respective to the distance between the electrodes relying on the selected research literature (Tables 1–3) where the Ohmic values are available.

Artery/Parameter	Radial	Ulnar	Brachial	Carotid	Tibial
max $\Delta Z(t)$ (Ω)	1.703	0.141	0.23	0.879	0.12
min $\Delta Z(t)$ (Ω)	0.545	0.084	0.038	0.303	0.03

The $\Delta Z(t)$ tends to be larger when acquired from the wrist—especially when targeted to the radial artery. Simultaneously, the $\Delta Z(t)$ is the lowest exclusively when acquired from the ulnar and tibial arteries.

5. Discussion

The presentation of sources analyzed in this review—namely research papers (Tables 1–3) and patents (Table 4)—differ substantially, making comprehensive and direct comparison challenging. In research papers, electrode designs and configurations are typically described in considerable detail. In contrast, patents often do not specify exact electrode dimensions or materials, even in implementation examples. Instead, they intentionally describe a wide range of possible options and configurations to cover the broadest possible scope of solutions in the claims.

https://doi.org/10.3390/instruments10020026

Accordingly, in the case of research papers, comparisons of EBI measurement results for targeted arteries can be performed based on the published data and the corresponding electrode arrangements. In contrast, patent descriptions are primarily useful for evaluating the proposed applications and measurement solutions for EBI, rather than for comparing quantitative results.

A clear indicator of the performance of different arteries in terms of impedance waveform quality is the magnitude of modulation, $\Delta Z(t)$, caused by the pulsatile blood volume. In practice, $\Delta Z(t)$ reflects measurement sensitivity—that is, how distinctly the modulation due to pulsatile blood flow (and other events within the cardiac cycle) can be extracted from the impedance waveform. Naturally, measurement techniques differ, electrode configurations vary, and data presentation methods are not uniform; nevertheless, the literature allows for qualitative assessment of which arteries provide higher sensitivity. As a reference, ref. [112] represents the first report to present a comprehensive analysis and comparison of impedance waveforms derived from five major peripheral arteries.

The value of $\Delta Z(t)$ is often not reported in the literature, as it is generally considered to be of limited informational value. Nevertheless, reporting the overall Ohmic values can provide useful insight, for example, into the quality of the skin–electrode contact interface. Numerical values of $\Delta Z(t)$ were reported in 16 of the evaluated research papers [19,30,45,49,50,68,69,74,75,77,80–82,86].

The radial artery, being the most popular and intuitive location for non-invasive pulse wave detection using EBI measurements, has been the most extensively studied in the literature [19,30,45,49,50,69–75,78,79,105]. The reported values of $\Delta Z(t)$ vary noticeably, but not significantly, largely depending on the selected electrode type and the exact electrode arrangement. Based on these reported values, a clear trend suggests the existence of an optimal electrode strategy, determined by electrode size and type in relation to the volume encompassed between the electrodes. Notably, similar observations apply to all targeted arteries discussed in the reviewed literature (Tables 1–3).

However, in a number of publications, measured values of $\Delta Z(t)$ are not reported. Instead, extracted features are presented and compared directly, such as PTT in [78], or the results of impedance measurements are reported in terms of measured voltages [71,72,113]. Nevertheless, for comparative purposes, trends related to electrode arrangement can still be evaluated across studies.

The chosen arterial sites are not exclusive, as the research, which in many cases is heavily commercialization-targeted, considers the device design from a usability perspective. For this reason, the fingers have also gained attention in EBI-based pulse wave signal acquisition due to their traditional location for rings. While PPG-based smart rings are today commercially available, supported by other sensorics via signal fusion [114], the EBI is still in its development phase [115]. Another more opportunistic location is the toe for possible acquisition of hemodynamic properties [116].

Placement of electrodes along the artery is a widely adopted approach, following the assumption that the pulsatile blood volume within the artery contributes dominantly to the measured EBI signal. Consequently, "in a row along the artery" configurations are predominant—accounting for approximately 95 % of the cases reported in Tables 1–3. However, alternative configurations also exist, such as electrode placement across the artery [78,80], supporting the hypothesis regarding current distribution within the volume conductor discussed in Section 1.

The four-electrode configuration is the predominant approach in pulse wave signal acquisition. This is primarily due to its advantage over the two-electrode configuration, namely the reduced influence of electrode impedances on the measured signal. Such a feature enables the detection of relatively small impedance variations (modulation depth)

https://doi.org/10.3390/instruments10020026

associated with cardiac activity. However, the two-electrode configuration offers benefits in terms of simpler hardware implementation. Furthermore, FEM simulations have demonstrated the presence of regions with positive sensitivity to arterial blood volume even in the two-electrode configuration [109].

With respect to the patents reviewed (Table 4), a greater variability of electrode arrangements has been proposed. However, in patent disclosures, a broad range of possible electrode configurations is typically described in order to protect the widest possible scope of applications. As a result, a larger number of targeted arteries may be defined, often with only general descriptions of electrode properties rather than detailed specifications.

Furthermore, patent documents usually do not describe a study population—unless they are based on an associated publication—and typically do not report quantitative Ohmic measurement values ($\Delta Z(t)$).

Clearly, the choice of electrode placement strategy is directly related to the targeted feature or, more broadly, to a specific pathology or medical condition. If the aim is to derive the PWV, the selection of the measurement site is relatively flexible, as the PWV has been obtained from the radial artery [49,71,73], the whole hand [65], and more classical sites such as the carotid and femoral arteries [84]. However, from a practical implementation perspective, the choice of artery or body region remains of substantial importance.

6. Conclusions, Challenges, and Opportunities

Considering current trends in EBI-based pulse wave detection, an additional extended arterial site can be identified—the palmar digital arteries supplying the fingers [115] and the toe digital artery [116]. Targeting these digital arteries by integrating electrodes into ring-shaped wearable devices is gaining popularity [117,118]. In such cases, miniaturization imposes increased demands on electrode configuration, leading to unconventional solutions based on four-electrode systems. One such solution is a rectangular four-electrode configuration (Figure 5c) [30].

Currently, there are no widely recognized mainstream commercial wearables based on RICG that use EBI alone for heart rate detection. Notable examples from the past include the Up3 and Up4 devices by Jawbone; however, these products were withdrawn from the market in 2017. Nevertheless, as indicated in Tables 1–4, this remains a highly promising research area with significant potential for innovation and patenting.

The $\Delta Z(t)$ component of the acquired impedance signal generally exhibits a very small magnitude, with some exceptions. In general, the farther the measurement site is from the heart, the smaller the observed $\Delta Z(t)$. The smaller the waveform variation caused by a physiological (e.g., cardiac) event, the higher the required signal quality—that is, a higher SNR.

Moreover, extensive signal processing demands substantial computational power, which in turn increases energy consumption. Although alternative approaches, such as analog feature extraction at the sensor edge [119], have been proposed, the primary objective should remain the acquisition of the highest possible signal quality at the source.

The most popular EBI-based pulse wave acquisition site is the hand, targeting the radial, ulnar, and brachial arteries (e.g., [40]). However, increasing attention has also been directed toward the leg [45,67,120,121], where the tibial artery reaches the skin surface and is suitable for targeting with miniaturized electrodes. In future solutions, electrodes—such as those made from conductive textiles—may be integrated into wearable items like socks or bands [122].

Advances in conductive textile materials are opening new horizons for EBI-based pulse wave acquisition, as discussed, for example, in [123,124], through integration into stretchable fabrics and e-textiles [125]. Conductive inks and screen-printing techniques on

https://doi.org/10.3390/instruments10020026

textiles already provide excellent electrical properties [126], potentially paving the way toward practical and usable EBI-based wearable devices [127].

In several cases reported in the reviewed literature (Tables 1–3), commercial wet gel electrodes have been used for EBI-based pulse wave signal acquisition (e.g., [30,40,45,68,75,80,82,84,105]). For research purposes, the use of commercial wet gel electrodes is reasonable, as they provide stable electrode–skin contact impedance across the frequency range typically employed in EBI measurements [31]. However, for commercial wearable applications, gel-based electrodes are impractical, and dry electrode solutions must be considered, despite their tendency to exhibit less stable contact impedance and frequency response.

When considering dry electrodes, a variety of additional factors must be addressed, including material choice, mechanical rigidity, and surface finish. Noble metals, such as silver [72,78] and gold [19], are generally safe options due to their biocompatibility, whereas copper [50] can be potentially allergenic. One proposed solution is the incorporation of highly conductive and solvent-resistant materials, such as carbon nanomaterials, into the electrode composition [31,128].

The optimization of electrode configurations is a key factor in achieving an EBI signal of the best possible quality. For signal waveform analysis, the original shape of the signal, representing the volume changes in the object, is expected to be present. There are two broad domains in which this can be achieved: hardware (1) and software (2). In configuring the electrodes for the most optimal setup via considering the peculiarities of the targeted arterial site, the hardware domain can be influenced. This includes additional sensors like accelerometers [36] or signal fusion in hemodynamic feature point detection [129]. The software domain—more precisely the signal processing task—can only perform as well as possible by relying on the originally acquired signal [20,130]. The classical signal processing methods are highly applicable, but today are extensively supported by artificial intelligence-based tools [131].

Machine learning (ML) methods provide new tools for hemodynamic feature extraction, particularly in challenging scenarios involving potentially noisy EBI signals [132]. However, because EBI methods are still rarely applied commercially for pulse wave acquisition and are largely confined to research settings, this field remains relatively understudied [133,134]. One primary reason for the limited number of studies applying ML methods is the specificity of EBI measurements, as the volume of acquired data is typically limited. ML approaches require substantial datasets for training; however, non-invasive EBI measurements are highly susceptible to motion artifacts, which further reduce the amount of usable data. In several cases, EBI signal acquisition has been performed on dedicated control groups to generate datasets for ML algorithm development [135]. Nevertheless, given the strong potential of EBI methods, continued research efforts are expected to expand available datasets and further advance ML-based approaches.

Funding: This research was supported by the Estonian Research Council Grant PRG1483 and co-funded by the European Union via the Astra+ 2021-2027.1.01.25-1170 project. The research was also funded by the European Union, and co-funded by the Ministry of Education and Research (Estonia) via the TEM-TA43 project.

Data Availability Statement: No new data were created or analyzed in this study. Data sharing is not applicable to this article.

Acknowledgments: The author's sincere thanks go to their colleagues at the Thomas Johann Seebeck Department of Electronics for the encouragement in compiling this review. During the preparation of this manuscript, the author used ChatGPT 5.2 for the purposes of text editing and grammar

https://doi.org/10.3390/instruments10020026

corrections. The authors have reviewed and edited the output and take full responsibility for the content of this publication.

Conflicts of Interest: The author declares no conflicts of interest. The funders had no role in the preparation of the manuscript or in the decision to publish the results.

Abbreviations

The following abbreviations are used in this manuscript:

EBI	Electrical Bioimpedance
ICG	Impedance Cardiography
RICG	Regional Impedance Cardiography
TICG	Thoracic Impedance Cardiography
WICG	Whole-Body Impedance Cardiography
ECG	Electrocardiogram
PEP	Pre-Ejection Period
RBC	Red Blood Cell
FEM	Finite Element Method
LVET	Left Ventricular Ejection Time
PWV	Pulse Wave Velocity
cfPWV	Carotid-Femoral Pulse Wave Velocity
BP	Blood Pressure
VTT	Vascular Transit Time
HR	Heart Rate
FIM	Focused Impedance Method
SNR	Signal-to-Noise Ratio
ML	Machine Learning
PTT	Pulse Transit Time

References

1. World Health Organization. Cardiovascular Diseases (CVDs). 2025. Available online: https://www.who.int/news-room/fact-sheets/detail/cardiovascular-diseases-(cvds) (accessed on 26 December 2025).
2. Lapsa, D.; Janeliukstis, R.; Metshein, M.; Selavo, L. PPG and Bioimpedance-Based Wearable Applications in Heart Rate Monitoring—A Comprehensive Review. *Appl. Sci.* **2024**, *14*, 7451. [CrossRef]
3. Jin, J.; Geng, X.; Zhang, Y.; Zhang, H.; Ye, T. Pulse Wave Analysis Method of Cardiovascular Parameters Extraction for Health Monitoring. *Int. J. Environ. Res. Public Health* **2023**, *20*, 2597. [CrossRef] [PubMed]
4. Arabsalmani, N.; Ghouchani, A.; Jafarabadi Ashtiani, S.; Zamani, M. Exploring Bio-Impedance Sensing for Intelligent Wearable Devices. *Bioengineering* **2025**, *12*, 521. [CrossRef]
5. Naranjo-Hernández, D.; Reina-Tosina, J.; Min, M. Fundamentals, Recent Advances, and Future Challenges in Bioimpedance Devices for Healthcare Applications. *J. Sens.* **2019**, *2019*, 9210258. [CrossRef]
6. Showkat, I.; Khanday, F.; Beigh, M. A review of bio-impedance devices. *Med. Biol. Eng. Comput.* **2023**, *61*, 927–950. [CrossRef]
7. Xu, J.; Hong, Z. Low Power Bio-Impedance Sensor Interfaces: Review and Electronics Design Methodology. *IEEE Rev. Biomed. Eng.* **2022**, *15*, 23–35. [CrossRef]
8. Mao, P.; Li, H.; Yu, Z. A Review of Skin-Wearable Sensors for Non-Invasive Health Monitoring Applications. *Sensors* **2023**, *23*, 3673. [CrossRef]
9. Vuorela, T.; Kukkonen, K.; Rantanen, J.; Jarvinen, T.; Vanhala, J. Bioimpedance measurement system for smart clothing. In Proceedings of the Seventh IEEE International Symposium on Wearable Computers, 2003, Proceedings, White Plains, NY, USA, 21–23 October 2003; pp. 98–107. [CrossRef]
10. Ferrari, R. Writing narrative style literature reviews. *Med. Writ.* **2015**, *24*, 230–235. [CrossRef]
11. Martinsen, O.G.; Grimnes, S. *Bioimpedance and Bioelectricity Basics*, 3rd ed.; Academic Press: Cambridge, MA, USA, 2014.
12. Metshein, M. Wearable Solutions for Monitoring Cardiorespiratory Activity. Ph.D. Dissertation, Tallinn University of Technology, Tallinn, Estonia, 2018.
13. Patterson, R. Fundamentals of impedance cardiography. *IEEE Eng. Med. Biol. Mag.* **1989**, *8*, 35–38. [CrossRef]
14. Sherwood, A.; Allen, M.T.; Fahrenberg, J.; Kelsey, R.M.; Lovallo, W.R.; van Doornen, L.J. Methodological Guidelines for Impedance Cardiography. *Psychophysiology* **1990**, *27*, 1–23. [CrossRef]

https://doi.org/10.3390/instruments10020026

15. Metshein, M.; Krivošei, A.; Abdullayev, A.; Annus, P.; Märtens, O. Non-Standard Electrode Placement Strategies for ECG Signal Acquisition. *Sensors* **2022**, *22*, 9351. [CrossRef]
16. Cybulski, G.; Strasz, A.; Niewiadomski, W.; Gąsiorowska, A. Impedance cardiography: Recent advancements. *Cardiol. J.* **2012**, *5*, 550–556. [CrossRef] [PubMed]
17. Abdelhammed, A.I.; Smith, R.D.; Levy, P.; Smits, G.J.; Ferrario, C.M. Noninvasive hemodynamic profiles in hypertensive subjects. *Am. J. Hypertens.* **2005**, *18*, 51S–59S. [CrossRef]
18. Ferrario, C.M.; Basile, J.; Bestermann, W.; Frohlich, E.; Houston, M.; Lackland, D.T.; Smith, R.D.; Wise, D.L. Review: The role of noninvasive hemodynamic monitoring in the evaluation and treatment of hypertension. *Ther. Adv. Cardiovasc. Dis.* **2007**, *1*, 113–118. [CrossRef] [PubMed]
19. Metshein, M.; Annus, P.; Land, R.; Min, M.; Aabloo, A. Availability and variations of cardiac activity in the case of measuring the bioimpedance of wrist. In Proceedings of the 2018 IEEE International Instrumentation and Measurement Technology Conference (I2MTC), Houston, TX, USA, 14–17 May 2018; pp. 1–5. [CrossRef]
20. Lapsa, D.; Metshein, M.; Krivošei, A.; Janeliukstis, R.; Märtens, O.; Elsts, A. Signal Acquisition and Algorithm Design for Bioimpedance-Based Heart Rate Estimation from the Wrist. *Appl. Sci.* **2024**, *14*, 9632. [CrossRef]
21. Andreuccetti, D.; Fossi, R.; Petrucci, C. *An Internet Resource for the Calculation of the Dielectric Properties of Body Tissues in the Frequency Range 10 Hz–100 GHz*; IFAC-CNR: Florence, Italy, 1997; Vol. Based on data published by C. Gabriel et al. in 1996 . Available online: http://niremf.ifac.cnr.it/tissprop/ (accessed on 29 December 2025).
22. Metshein, M.; Annus, P.; Land, R.; Rist, M.; Min, M.; Lotamõis, K.; Märtens, O. On-Desk Model-Based Study for Evaluating the Effect of Squeezing the Arteries in the Forearm on the Bioimpedance and Pressure of Blood. *J. Sens.* **2023**, *2023*, 5710835. [CrossRef]
23. Sigman, E.; Kolin, A.; Katz, L.N.; Jochim, K. Effect of motion on the electrical conductivity of the blood. *Am. J. Physiol.-Leg. Content* **1937**, *118*, 708–719. [CrossRef]
24. Yamakoshi, K.I.; Shimazu, H.; Togawa, T.; Fukuoka, M.; Ito, H. Noninvasive Measurement of Hematocrit by Electrical Admittance Plethysmography Technique. *IEEE Trans. Biomed. Eng.* **1980**, *BME-27*, 156–161. [CrossRef]
25. McAdams, E. Bioelectrodes. In *Encyclopedia of Medical Devices and Instrumentation, Set*; John Wiley & Sons, Ltd.: Hoboken, NJ, USA, 2006. [CrossRef]
26. Kauppinen, P.; Hyttinen, J.; Malmivuo, J. Sensitivity Distributions of Impedance Cardiography Using Band and Spot Electrodes Analyzed by a Three-Dimensional Computer Model. *Ann. Biomed. Eng.* **1998**, *26*, 694–702. [CrossRef]
27. Patterson, R.P. Impedance cardiography: What is the source of the signal? *J. Phys. Conf. Ser.* **2010**, *224*, 012118. [CrossRef]
28. Sramek, B. Status report on BoMed's electrical bioimpedance. In Proceedings of the Annual International Conference of the IEEE Engineering in Medicine and Biology Society, New Orleans, LA, USA, 4–7 November 1988; Volume 1, p. 51. [CrossRef]
29. Kubicek, W.G.; Karnegis, J.N.; Patterson, R.P.; Witsoe, D.A.; Mattson, R.H. Development and evaluation of an impedance cardiac output system. *Aerosp. Med. Hum. Perform.* **1966**, *37*, 1208–1212.
30. Pesti, K.; Metshein, M.; Annus, P.; Kõiv, H.; Min, M. Electrode Placement Strategies for the Measurement of Radial Artery Bioimpedance: Simulations and Experiments. *IEEE Trans. Instrum. Meas.* **2021**, *70*, 9500610. [CrossRef]
31. Yang, L.; Gan, L.; Zhang, Z.; Zhang, Z.; Yang, H.; Zhang, Y.; Wu, J. Insight into the Contact Impedance between the Electrode and the Skin Surface for Electrophysical Recordings. *ACS Omega* **2022**, *7*, 13906–13912. [CrossRef]
32. Marcôndes, D.W.C.; Paterno, A.S.; Bertemes-Filho, P. Parasitic Effects on Electrical Bioimpedance Systems: Critical Review. *Sensors* **2022**, *22*, 8705. [CrossRef]
33. Hafid, A.; Zolfaghari, S.; Kristoffersson, A.; Folke, M. Exploring the potential of electrical bioimpedance technique for analyzing physical activity. *Front. Physiol.* **2024**, *15*, 1515431. [CrossRef]
34. Cömert, A.; Hyttinen, J. Impedance spectroscopy of changes in skin-electrode impedance induced by motion. *BioMed. Eng. OnLine* **2014**, *13*, 149. [CrossRef]
35. An, X.; Liu, Y.; Zhao, Y.; Lu, S.; Stylios, G.K.; Liu, Q. Adaptive Motion Artifact Reduction in Wearable ECG Measurements Using Impedance Pneumography Signal. *Sensors* **2022**, *22*, 5493. [CrossRef]
36. Dang, T.H.; Jang, G.Y.; Lee, K.; Oh, T.I. Motion Artifacts Reduction for Noninvasive Hemodynamic Monitoring of Conscious Patients Using Electrical Impedance Tomography: A Preliminary Study. *Sensors* **2023**, *23*, 5308. [CrossRef] [PubMed]
37. Gruetzmann, A.; Hansen, S.; Müller, J. Novel dry electrodes for ECG monitoring. *Physiol. Meas.* **2007**, *28*, 1375. [CrossRef] [PubMed]
38. Kassanos, P. Bioimpedance Sensors: A Tutorial. *IEEE Sens. J.* **2021**, *21*, 22190–22219. [CrossRef]
39. Martinsen, O.G.; Hou, J. Modeling two- and four-electrode impedance measurements on a layered material. In Proceedings of the 2024 19th Biennial Baltic Electronics Conference (BEC), Tallinn, Estonia, 2–4 October 2024; pp. 1–4. [CrossRef]

https://doi.org/10.3390/instruments10020026

40. Xu, J.; Gao, X.; Lee, A.; Yamada, S.; Yavari, E.; Lubecke, V.; Boric-Lubecke, O. Wrist-worn heartbeat monitoring system based on bio-impedance analysis. In Proceedings of the 2016 38th Annual International Conference of the IEEE Engineering in Medicine and Biology Society (EMBC), Orlando, FL, USA, 16–20 August 2016; pp. 6294–6297. [CrossRef]
41. Sola, J.; Adler, A.; Santos, A.; Tusman, G.; Sipmann, F.; Bohm, S. Non-invasive monitoring of central blood pressure by electrical impedance tomography: First experimental evidence. *Med. Biol. Eng. Comput.* **2011**, *49*, 409–415. [CrossRef]
42. Müller-Graf, F.; Thönes, J.; Krukewitt, L.; Frenkel, P.; Richter, H.; Spors, S.; Kühn, V.; Zitzmann, A.; Boehm, S.; Reuter, D. Non-invasive estimation of beat-by-beat aortic blood pressures from electrical impedance tomography data processed by machine learning. *J. Clin. Monit. Comput.* **2025**, *39*, 841–852. [CrossRef]
43. Han, T.; Qin, Y.; Zhao, Z.; Yang, B.; Liu, X.; Li, L.; Wei, Z.; Wei, L.; Liu, Y.; Fu, F. Calibration of ventilation/perfusion match in electrical impedance tomography: A novel method based on arterial blood pressure. *Front. Physiol.* **2025**, *16*, 1545652. [CrossRef]
44. Brandao, A.A.; Amodeo, C.; Alcantara, C.; Barbosa, E.; Nobre, F.; Pinto, F.; Vilela-Martin, J.F.; Bastos, J.M.; Yugar-Toledo, J.C.; Mota-Gomes, M.A.; et al. I Luso-Brazilian Positioning on Central Arterial Pressure. *Arq. Bras. Cardiol.* **2017**, *108*, 100–108. [CrossRef] [PubMed]
45. Sel, K.; Osman, D.; Jafari, R. Non-Invasive Cardiac and Respiratory Activity Assessment from Various Human Body Locations Using Bioimpedance. *IEEE Open J. Eng. Med. Biol.* **2021**, *2*, 210–217. [CrossRef] [PubMed]
46. Nichols, W.W.; Edwards, D.G. Arterial Elastance and Wave Reflection Augmentation of Systolic Blood Pressure: Deleterious Effects and Implications for Therapy. *J. Cardiovasc. Pharmacol. Ther.* **2001**, *6*, 5–21. [CrossRef] [PubMed]
47. Wang, J.J.; Wang, P.W.; Liu, C.P.; Lin, S.K.; Hu, W.C.; Kao, T. Evaluation of changes in cardiac output from the electrical impedance waveform in the forearm. *Physiol. Meas.* **2007**, *28*, 989–999. [CrossRef]
48. Anderson, F.A. Impedance plethysmography in the diagnosis of arterial and venous disease. *Ann. Biomed. Eng.* **2006**, *12*, 79–102. [CrossRef]
49. Cho, M.C.; Kim, J.Y.; Cho, S.H. A bio-impedance measurement system for portable monitoring of heart rate and pulse wave velocity using small body area. In Proceedings of the 2009 IEEE International Symposium on Circuits and Systems (ISCAS), Taipei, Taiwan, 24–27 May 2009; pp. 3106–3109. [CrossRef]
50. Huynh, T.H.; Jafari, R.; Chung, W.Y. An Accurate Bioimpedance Measurement System for Blood Pressure Monitoring. *Sensors* **2018**, *18*, 2095. [CrossRef]
51. Aatola, H.; Hutri-Kähönen, N.; Juonala, M.; Viikari, J.S.; Hulkkonen, J.; Laitinen, T.; Taittonen, L.; Lehtimäki, T.; Raitakari, O.T.; Kähönen, M. Lifetime Risk Factors and Arterial Pulse Wave Velocity in Adulthood. *Hypertension* **2010**, *55*, 806–811. [CrossRef]
52. Cotter, G.; Moshkovitz, Y.; Kaluski, E.; Cohen, A.J.; Miller, H.; Goor, D.A.; Vered, Z. Accurate, noninvasive continuous monitoring of cardiac output by whole-body electrical bioimpedance. *Chest* **2004**, *125*, 1431–1440. [CrossRef]
53. Rabbani, K.S.; Sarker, M.; Akond, M.H.R.; Aakter, T. Focused Impedance Measurement (FIM): A New Technique with Improved Zone Localization. *Ann. N. Y. Acad. Sci.* **1999**, *873*, 408–420. [CrossRef]
54. Hai, V.D.; Hung, P.D.; Dan, C.Q. Modified electrode placements for measurement of hemodynamic parameters using impedance cardiography. *J. Med. Eng. Technol.* **2020**, *44*, 357–367. [CrossRef] [PubMed]
55. Nyboer, J.; Kreider, M.M.; Hannapel, L. Electrical Impedance Plethysmography. *Circulation* **1950**, *2*, 811–821. [CrossRef]
56. Vedru, J. Electrical impedance methods for the measurement of stroke volume in man: State of art. *Acta Comment. Univ. Tartu.* **1994**, *974*, 110–129 .
57. Austad., H.O.; Vedum., J.; Røed., M.H.; Dalgard., S.; Brødreskift., T.; Liverud., A.E.; Strisland., F.; Seeberg, T.M. An Unobtrusive Wearable Device for Ambulatory Monitoring of Pulse Transit Time to Estimate Central Blood Pressure. In *Proceedings of the 9th International Joint Conference on Biomedical Engineering Systems and Technologies—BIODEVICES, (BIOSTEC 2016), INSTICC, Rome, Italy, 21–23 February 2016*; SciTePress: Setubal, Portugal, 2016; pp. 179–186. [CrossRef]
58. Metshein, M.; Gautier, A.; Larras, B.; Frappe, A.; John, D.; Cardiff, B.; Annus, P.; Land, R.; Martens, O. Study of Electrode Locations for Joint Acquisition of Impedance- and Electro-cardiography Signals. In Proceedings of the 2021 43rd Annual International Conference of the IEEE Engineering in Medicine & Biology Society (EMBC), Guadalajara, Mexico, 1–5 November 2021; p. 7264. [CrossRef]
59. Hafid, A.; Benouar, S.; Kedir-Talha, M.; Attari, M.; Seoane, F. Simultaneous Recording of ICG and ECG Using Z-RPI Device with Minimum Number of Electrodes. *J. Sens.* **2018**, *2018*, 3269534:1–3269534:7. [CrossRef]
60. Metshein, M.; Tuulik, V.R.; Tuulik, V.; Min, M.; Kumm, M.; Annus, P.; Märtens, O. Body Fat Proportion Estimation by the Segmental Bioelectrical Impedance Analysis of Forearm. In Proceedings of the 2024 IEEE International Instrumentation and Measurement Technology Conference (I2MTC), Glasgow, UK, 20–23 May 2024; pp. 1–5. [CrossRef]
61. Liebich, C.; Bartsch, J.N.; Schubert, I.; von Bruehl, M.L.; Sander, C. Electrical Impedance Spectroscopy Improves Skin Cancer Detection and Reduces the Number of Biopsies. *Dermato* **2022**, *2*, 21–29. [CrossRef]

https://doi.org/10.3390/instruments10020026

62. Millasseau, S.C.; Stewart, A.D.; Patel, S.J.; Redwood, S.R.; Chowienczyk, P.J. Evaluation of Carotid–Femoral Pulse Wave Velocity: Influence of Timing Algorithm and Heart Rate. *Hypertension* **2005**, *45*, 222–226. [CrossRef] [PubMed]
63. Laurent, S.; Cockcroft, J.; Van Bortel, L.; Boutouyrie, P.; Giannattasio, C.; Hayoz, D.; Pannier, B.; Vlachopoulos, C.; Wilkinson, I.; Struijker-Boudier, H.; et al. Expert consensus document on arterial stiffness: Methodological issues and clinical applications. *Eur. Heart J.* **2006**, *27*, 2588–2605. [CrossRef]
64. Dzeko, M.; Peters, C.D.; Kjaergaard, K.D.; Jensen, J.D.; Jespersen, B. Aortic Pulse Wave Velocity Results Depend on Which Carotid Artery is Used for the Measurements. *J. Hypertens.* **2013**, *31*, 117–122. [CrossRef] [PubMed]
65. Aria, S.; Elfarri, Y.; Elvegård, M.; Gottfridsson, A.; Grønaas, H.S.; Harang, S.; Jansen, A.; Madland, T.E.R.; Martins, I.B.; Olstad, M.W.; et al. Measuring Blood Pulse Wave Velocity with Bioimpedance in Different Age Groups. *Sensors* **2019**, *19*, 850. [CrossRef]
66. Nabian, M.; Yin, Y.; Wormwood, J.; Quigley, K.S.; Barrett, L.F.; Ostadabbas, S. An Open-Source Feature Extraction Tool for the Analysis of Peripheral Physiological Data. *IEEE J. Transl. Eng. Health Med.* **2018**, *6*, 2800711. [CrossRef]
67. Metshein, M.; Abdullayev, A.; Gautier, A.; Larras, B.; Frappe, A.; Cardiff, B.; Annus, P.; Land, R.; Märtens, O. Sensor-Location-Specific Joint Acquisition of Peripheral Artery Bioimpedance and Photoplethysmogram for Wearable Applications. *Sensors* **2023**, *23*, 7111. [CrossRef]
68. Bernstein, D.P.; Henry, I.C.; Banet, M.J.; Dittrich, T. Stroke volume obtained by electrical interrogation of the brachial artery: Transbrachial electrical bioimpedance velocimetry. *Physiol. Meas.* **2012**, *33*, 629–649. [CrossRef]
69. Namkoong, M.; Baskar, B.; Singh, L.; Guo, H.; McMurray, J.; Branan, K.; Rahman, M.S.; Hsiao, C.T.; Kuriakose, J.; Hernandez, J.; et al. Add-On Soft Electronic Interfaces for Continuous Cuffless Blood Pressure Monitoring. *Adv. Mater. Technol.* **2023**, *8*, 2300158 . [CrossRef]
70. Namkoong, M.; McMurray, J.; Branan, K.; Hernandez, J.; Gandhi, M.; Ida-Oze, S.; Cote, G.; Tian, L. Contact Pressure-Guided Wearable Dual-Channel Bioimpedance Device for Continuous Hemodynamic Monitoring. *Adv. Mater. Technol.* **2024**, *9*, 2301407 . [CrossRef]
71. Huang, J.J.; Huang, Y.M.; Chang, M.W. Using bioimpedance plethysmography for measuring the pulse wave velocity of peripheral vascular. In Proceedings of the 2016 13th International Conference on Electrical Engineering/Electronics, Computer, Telecommunications and Information Technology (ECTI-CON), Chiang Mai, Thailand, 28 June 2016–1 July 2016; pp. 1–5. [CrossRef]
72. Huynh, T.H.; Jafari, R.; Chung, W.Y. Noninvasive Cuffless Blood Pressure Estimation Using Pulse Transit Time and Impedance Plethysmography. *IEEE Trans. Biomed. Eng.* **2019**, *66*, 967–976. [CrossRef]
73. Lee, K.; Yoo, H.J. Simultaneous Electrical Bio-Impedance Plethysmography at Different Body Parts: Continuous and Non-Invasive Monitoring of Pulse Wave Velocity. *IEEE Trans. Biomed. Circuits Syst.* **2021**, *15*, 1027–1038. [CrossRef]
74. Ibrahim, B.; Jafari, R. Cuffless Blood Pressure Monitoring From a Wristband with Calibration-free Algorithms for Sensing Location Based on Bio-impedance Sensor Array and Autoencoder. *IEEE Trans. Biomed. Circuits Syst.* **2022**, *12*, 319. [CrossRef]
75. Anand, G.; Lowe, A. Investigating Electrical Impedance Spectroscopy for Estimating Blood Flow-Induced Variations in Human Forearm. *Sensors* **2020**, *20*, 5333. [CrossRef]
76. Ibrahim, B.; Mrugala, D.; Jafari, R. Effects of Bio-Impedance Sensor Placement Relative to the Arterial Sites for Capturing Hemodynamic Parameters. In Proceedings of the 2019 41st Annual International Conference of the IEEE Engineering in Medicine and Biology Society (EMBC), Berlin, Germany, 23–27 July 2019; pp. 6569–6573. [CrossRef]
77. Imbriglia, M.; Picariello, F.; Tudosa, I.; De Vito, L.; Daponte, P. Assessment of a Bioimpedance Analog Front-End for Blood Pulse Wave Detection. In Proceedings of the 2024 IEEE International Symposium on Medical Measurements and Applications (MeMeA), Eindhoven, The Netherlands, 26–28 June 2024; pp. 1–6. [CrossRef]
78. Rachim, V.; Chung, W. Multimodal Wrist Biosensor for Wearable Cuff-less Blood Pressure Monitoring System. *Sci. Rep.* **2019**, *9*, 7947. [CrossRef]
79. Bang, S.; Lee, C.; Park, J.; Cho, M.C.; Yoon, Y.G.; Cho, S. A pulse transit time measurement method based on electrocardiography and bioimpedance. In Proceedings of the 2009 IEEE Biomedical Circuits and Systems Conference, Beijing, China, 26–28 November 2009; pp. 153–156. [CrossRef]
80. Liu, S.H.; Cheng, D.C.; Su, C.H. A Cuffless Blood Pressure Measurement Based on the Impedance Plethysmography Technique. *Sensors* **2017**, *17*, 1176. [CrossRef]
81. Ibrahim, B.; Hall, D.A.; Jafari, R. Bio-impedance spectroscopy (BIS) measurement system for wearable devices. In Proceedings of the 2017 IEEE Biomedical Circuits and Systems Conference (BioCAS), Turin, Italy, 19–21 October 2017; pp. 1–4. [CrossRef]
82. Wang, T.W.; Chu, H.W.; Chou, L.; Sung, Y.L.; Shih, Y.T.; Hsu, P.C.; Cheng, H.M.; Lin, S.F. Bio-Impedance Measurement Optimization for High-Resolution Carotid Pulse Sensing. *Sensors* **2021**, *21*, 1600. [CrossRef]
83. Wang, T.W.; Syu, J.Y.; Chu, H.W.; Sung, Y.L.; Chou, L.; Escott, E.; Escott, O.; Lin, T.T.; Lin, S.F. Intelligent Bio-Impedance System for Personalized Continuous Blood Pressure Measurement. *Biosensors* **2022**, *12*, 150. [CrossRef]
84. Collette, M.; Humeau, A.; Chevalier, C.; Hamel, J.F.; Georges, L. Assessment of Aortic Stiffness by Local and Regional Methods. *Hypertens. Res.* **2011**, *34*, 578–583. [CrossRef]

85. Diaz, D.H.; Casas, O.; Pallas-Areny, R. Heart rate detection from single-foot plantar bioimpedance measurements in a weighing scale. In Proceedings of the 2010 Annual International Conference of the IEEE Engineering in Medicine and Biology, Buenos Aires, Argentina, 31 August 2010–4 September 2010; pp. 6489–6492. [CrossRef]
86. Tronstad, C.; Callender, N.A.; Badeli, V.; Hisdal, J. A Flexible Electrode Strap for Impedance Plethysmography of the Dorsalis Pedis Artery. In Proceedings of the 2023 IEEE International Workshop on Metrology for Industry 4.0 & IoT (MetroInd4.0&IoT), Brescia, Italy, 6–8 June 2023; pp. 113–117. [CrossRef]
87. Tran, B. Smart Watch. US Patent 10 610 111, 7 April 2020.
88. Cohn, G.A.; Kusche, R. Bioimpedance Based Pulse Waveform Sensing. US Patent Application No. 15/620,651, 22 September 2016; abandoned.
89. Trapero, M.A. Electronic Device for Determining Bioimpedance. US Patent 11 320 789, 3 May 2022.
90. Gorbunov, S.; Dorokhin, I.; Davydov, D.; Boev, A. Smart Watch Strap for Bioimpedance Measurements. US Patent 11 185 246, 30 November 2021.
91. Podhajsky, R.; Reschke, A.J. Measuring Tissue Volume with Dynamic Autoreconfiguration. US Patent Application No. 14/491,836, 19 September 2014; abandoned.
92. Lading, L.; Baek, D.B. Method for Determining Pulse Wave Velocity in an Artery. US Patent 10 959 622, 30 March 2021.
93. Vule, Y.; Aminov, Y.; Sromin, A.; Galeev, A.; Dickinson, E.; Gorev, N. Systems and Methods for Collecting Physiological Information of a User. US Patent 11 317 814, 3 May 2022.
94. Lading, L.; Bæk, D.B. Method of and a System for Determining a Cardiovascular Quantity of a Mammal. US Patent 9 603 533, 28 March 2017.
95. Banet, M.; Henry, I.; Bernstein, D. Body-Worn System for Continuous, Noninvasive Measurement of Cardiac Output, Stroke Volume, Cardiac Power, and Blood Pressure. US Patent 10 278 599, 7 May 2019.
96. Davis, R.; Morrill, N.; Miller, D. Impedance Sensors for Detecting Heart Wave Forms. US Patent 11 284 809, 29 March 2022.
97. Stump, K. Enhanced Physiological Monitoring Devices and Computer-Implemented Systems and Methods of Remote Physiological Monitoring of Subjects. US Patent Application No. 17/026,929, 21 September 2020.
98. Banet, W.; Pede, S.M.; Dhillon, M.S.; Terry, A.; Hunt, K.R. Necklace-Shaped Physiological Monitor. US Patent 9 211 073, 12 December 2015.
99. Banet, M.; Dhillon, M.S.; Pede, S.M.; Hayward, L.M.; Dhillon, M.S.; Klein, J.; Stainer, D.; Broadbooks, R.C. Physiological monitor for MONITORING Patients Undergoing Hemodialysis. US Patent Application No. 62/346,410, 6 June 2016.
100. Gopinathan, V.; Doscher, J.; Akl, T. Multi-Sensor Device for Monitoring Health. US Patent Application No. 16/478,848, 24 January 2018.
101. Böttrich, M.; Tanskanen, J.; Hyttinen, J. Lead field theory provides a powerful tool for designing microelectrode array impedance measurements for biological cell detection and observation. *Biomed. Eng. OnLine* **2017**, *16*, 85. [CrossRef]
102. Schneider, J.; Schroth, M.; Holzhey, M.; Blöcher, T.; Stork, W. An approach to improve impedance plethysmography on the wrist by using adaptive feedback control. In Proceedings of the 2017 IEEE Sensors Applications Symposium (SAS), Glassboro, NJ, USA, 13–15 March 2017; pp. 1–6. [CrossRef]
103. Ibrahim, B.; McMurray, J.; Jafari, R. A Wrist-Worn Strap with an Array of Electrodes for Robust Physiological Sensing. In Proceedings of the 2018 40th Annual International Conference of the IEEE Engineering in Medicine and Biology Society (EMBC), Honolulu, HI, USA, 18–21 July 2018; pp. 4313–4317. [CrossRef]
104. Siddique-e Rabbani, K. Focused Impedance Method (FIM) and Pigeon Hole Imaging (PHI) as two potentially low cost and simple modalities for different diagnostic applications. In Proceedings of the 7th International Conference on Appropriate Healthcare Technologies for Developing Countries, London, UK, 18–19 September 2012; pp. 1–6. [CrossRef]
105. Lee, W.; Cho, S. Integrated All Electrical Pulse Wave Velocity and Respiration Sensors Using Bio-Impedance. *IEEE J. Solid-State Circuits* **2015**, *50*, 776–785. [CrossRef]
106. Korthuis, R. Chapter 3: Regulation of Vascular Tone in Skeletal Muscle. In *Skeletal Muscle Circulation*, 1st ed.; Morgan and Claypool Life Sciences: San Rafael, CA, USA, 2011.
107. Vlachopoulos, C.; O'Rourke, M.; Nichols, W. *McDonald's Blood Flow in Arteries: Theoretical, Experimental and Clinical Principles*, 6th ed.; CRC Press: Boca Raton, FL, USA, 2011.
108. Jivet, I. Wrist pulse monitoring by electrical impedance using a 3D model of the arm. In Proceedings of the 2014 11th International Symposium on Electronics and Telecommunications (ISETC), Timisoara, Romania, 14–15 November 2014; pp. 1–4. [CrossRef]
109. Metshein, M.; Pesti, K.; Lapsa, D.; Annus, P.; Janeliukstis, R.; Elsts, A.; Martens, O. Evaluation of Two-Electrode System Configurations for Forearm Arteries Bioimpedance Measurement. In Proceedings of the 2024 IEEE International Instrumentation and Measurement Technology Conference (I2MTC), Glasgow, UK, 20–23 May 2024; pp. 1–6. [CrossRef]

https://doi.org/10.3390/instruments10020026

110. Jung, M.H.; Namkoong, K.; Lee, Y.; Koh, Y.J.; Eom, K.; Jang, H.; Bae, J.; Park, J. Wrist-wearable bioelectrical impedance analyzer with contact resistance compensation function. In Proceedings of the 2016 IEEE SENSORS, Orlando, FL, USA, 30 October 2016–3 November 2016; pp. 1–3. [CrossRef]
111. Bogónez-Franco, P.; Pham, P.; Géhin, C.; Massot, B.; Delhomme, G.; Guillemaud, R.; McAdams, E.T. Effect of electrode contact impedance mismatch on 4-electrode measurements of small body segments using commercial BIA devices. In Proceedings of the 20th IMEKO TC4 International Symposium and 18th International Workshop on ADC Modelling and Testing, Benevento, Italy, 15–17 September 2014; pp. 895–899.
112. Yazdanian, H.; Mahnam, A.; Edrisi, M.; Esfahani, M. Design and implementation of a portable impedance cardiography system for noninvasive stroke volume monitoring. *J. Med. Signals Sens.* **2016**, *6*, 47–56. [CrossRef]
113. Metshein, M.; Parve, T.; Annus, P.; Rist, M.; Min, M. Realization and Evaluation of the Device for Measuring the Impedance of Human Body for Detecting the Respiratory and Heart Rate. *Elektron. Elektrotechnika* **2017**, *3*, 36–42. [CrossRef]
114. Wang, M.; Shi, W.; Zhang, J.; Chen, J.; Lin, Q.; Chen, C.; Wang, G. Development of a Rotation-Robust PPG Sensor for a Smart Ring. *Sensors* **2025**, *25*, 6326. [CrossRef]
115. Sel, K.; Osman, D.; Huerta, N.; Edgar, A.; Pettigrew, R.I.; Jafari, R. Continuous cuffless blood pressure monitoring with a wearable ring bioimpedance device. *npj Digit. Med.* **2023**, *6*, 59. [CrossRef] [PubMed]
116. Kalvøy, H.; Hisdal, J.; Tronstad, C. Toe digit pulsation measurement by impedance plethysmography in miniaturized devices: An early feasibility test. *J. Electr. Bioimpedance* **2025**, *16*, 150–152. [CrossRef]
117. Ha, G.B.; Boyadzhiev, A.; Zhang, H.; Sanchez, B. A Wearable Smart Ring for Measuring Bioimpedance in the Palmar Digital Arteries. In Proceedings of the 2024 IEEE 20th International Conference on Body Sensor Networks (BSN), Chicago, IL, USA, 15–17 October 2024; pp. 1–4. [CrossRef]
118. Osman, D.; Jankovic, M.; Sel, K.; Pettigrew, R.I.; Jafari, R. Blood Pressure Estimation using a Single Channel Bio-Impedance Ring Sensor. In Proceedings of the 2022 44th Annual International Conference of the IEEE Engineering in Medicine & Biology Society (EMBC), Glasgow, UK, 11–15 July 2022; pp. 4286–4290. [CrossRef]
119. Martens, O.; Metshein, M.; Abdullayev, A.; Larras, B.; Frappe, A.; Gautier, A.; Saeed, M.; John, D.; Cardiff, B.; Krivošei, A.; et al. Fiducial Point Estimation Solution for Impedance Cardiography Measurements. In Proceedings of the 2022 IEEE International Instrumentation and Measurement Technology Conference (I2MTC), Ottawa, ON, Canada, 16–19 May 2022; pp. 1–6. [CrossRef]
120. Hong, S.; Coté, G. Development of a Tetherless Bioimpedance Device That Uses Morphologic Changes to Predict Blood Flow Restrictions Mimicking Peripheral Artery Disease Progression. *Biosensors* **2024**, *14*, 286. [CrossRef]
121. Horinaka, S.; Sakuma, M.; Yonezawa, Y.; Watahiki, M.; Higano, C.; Toyoda, S.; Yamamoto, T. Usefulness of Blood Flow Measurement Device Using Bioelectrical Impedance Plethysmography in Lower-Extremity Artery Disease. *Circ. Rep.* **2025**, *7*, 113–121. [CrossRef] [PubMed]
122. Sánchez, M.J.; Scagliusi, S.J.F.; Giménez-Miranda, L.; Pérez, P.; Medrano, F.J.; Olmo Fernández, A. Design of Wearable Textile Electrodes for the Monitorization of Patients with Heart Failure. *Sensors* **2024**, *24*, 3637. [CrossRef]
123. Marquez, J.; Rempfler, M.; Seoane, F.; Lindecrantz, K. Textrode-enabled transthoracic electrical bioimpedance measurements—Towards wearable applications of impedance cardiography. *J. Electr. Bioimpedance* **2013**, *4*, 45–50. [CrossRef]
124. Ravichandran, V.; Ciesielska-Wrobel, I.; Rumon, M.A.A.; Solanki, D.; Mankodiya, K. Characterizing the Impedance Properties of Dry E-Textile Electrodes Based on Contact Force and Perspiration. *Biosensors* **2023**, *13*, 728. [CrossRef]
125. Sharma, S.; Shukla, S.K.; Singh, A.; Govender, K.K.; Govender, P.P. E-Textiles in Biomedicine: Real Time Sensing, Energy Storage, and Therapeutic Applications. *Adv. Mater. Interfaces* **2026**, *13*, e00672. [CrossRef]
126. Kim, J.; Choi, H.; Kim, B.; Kim, E.; Kim, J. Flexible electrode on e-textile based on screen-printed silver ink carbon nanotube. *Int. J. Adv. Manuf. Technol.* **2024**, *134*, 127–137. [CrossRef]
127. Blachowicz, T.; Ehrmann, G.; Ehrmann, A. Textile-Based Sensors for Biosignal Detection and Monitoring. *Sensors* **2021**, *21*, 6042. [CrossRef]
128. Kõiv, H.; Pesti, K.; Min, M.; Land, R. Investigation of cost-effective carbon nanofiber/carbon fiber and silicone polymer composite material for wearable bioimpedance device. In Proceedings of the 2019 IEEE Sensors Applications Symposium (SAS), Sophia Antipolis, France, 11–13 March 2019; pp. 1–6. [CrossRef]
129. Metshein, M.; Annus, P.; Abdullayev, A.; Krivošei, A.; Land, R.; Märtens, O. Photoplethysmography as an Alternative to Impedance Cardiography-Based Hemodynamic Feature Point Detection. In Proceedings of the 2024 19th Biennial Baltic Electronics Conference (BEC), Tallinn, Estonia, 2–4 October 2024; pp. 1–5. [CrossRef]
130. Chabchoub, S.; Mansouri, S.; Salah, R.B. Signal processing techniques applied to impedance cardiography ICG signals—A review. *J. Med. Eng. Technol.* **2022**, *46*, 243–260. [CrossRef] [PubMed]
131. Jacobs, I.; Lowe, A.; Garcia, L.; Zhang, H. Advancing physical activity monitoring through bioimpedance measurement: A review. *Prog. Biomed. Eng.* **2026**, *8*, 022002. [CrossRef]
132. Liu, S.H.; Sun, Y.; Wu, B.Y.; Chen, W.; Zhu, X. Using machine learning models for cuffless blood pressure estimation with ballistocardiogram and impedance plethysmogram. *Front. Digit. Health* **2025**, *7*, 1511667. [CrossRef] [PubMed]

https://doi.org/10.3390/instruments10020026

133. Tronstad, C.; Strand-Amundsen, R. Possibilities in the application of machine learning on bioimpedance time-series. *J. Electr. Bioimpedance* **2019**, *10*, 24–33. [CrossRef] [PubMed]
134. Hou, J.; Nesaragi, N.; Tronstad, C. Electrical bioimpedance in the era of artificial intelligence. *J. Electr. Bioimpedance* **2024**, *15*, 1–3. [CrossRef] [PubMed]
135. Ames-Lastra, G.; Concu, A.; Real-Moreno, O.; Kataishi, R.E.; Dell'Osa, A.H. Evaluation of Machine Learning algorithms applied to differentiate upper extremities from bioimpedance measurements. In Proceedings of the 2023 20th International Conference on Electrical Engineering, Computing Science and Automatic Control (CCE), Mexico City, Mexico, 25–27 October 2023; pp. 1–5. [CrossRef]

https://doi.org/10.3390/instruments10020026

MDPI AG
Grosspeteranlage 5
4052 Basel
Switzerland
Tel.: +41 61 683 77 34

Instruments Editorial Office
E-mail: instruments@mdpi.com
www.mdpi.com/journal/instruments

www.ingramcontent.com/pod-product-compliance
Lightning Source LLC
Chambersburg PA
CBHW061629210926
51850CB00009B/2488
* 9 7 8 3 7 2 5 8 8 1 6 1 1 *